U.S.-UKRAINE MILITARY RELATIONS AND THE VALUE OF INTEROPERABILITY

Leonid I. Polyakov

December 2004

ACKNOWLEDGEMENTS

I am grateful to many people who supported this monograph with their knowledge and advice. First of all, I thank the authors of interviews for sharing their knowledge and experiences: U.S. Army retired Major General Nicholas Krawciw, U.S. Army retired Colonel Harry Simmeth, Ukrainian Army retired Colonel Dmitri Shkurko, Ukrainian Army Colonel Sergiy Poliakov, U.S. Army Colonel Timothy Shea, U.S. Air Force Lieutenant Colonel John Cappello, and California Army National Guard Major Joel Ostrom. This report would not be possible without friendly revision, skillful insight, and all possible support from my friends, Colonel Harry Simmeth and Mr. James Greene. Special thanks to my 1995 U.S. Army War College classmate, Colonel Harry Simmeth, for his expert opinion on the early days of U.S.-Ukraine military engagement. This came from his unique perspective of having been the principal U.S. architect of that engagement while serving as a U.S. Army colonel on the U.S. Joint Staff in the Pentagon. Important additions to the report were made due to the timely recommendations by U.S. Army Colonel Mark Easton and U.S. Army Lieutenant Colonel Leigh Gabel. I am grateful to the U.S. Army Strategic Studies Institute (SSI), in particular to the Director, Professor Douglas C. Lovelace, Jr., who encouraged me to write the monograph, and to Dr. Stephen Blank, for "piloting" it through the SSI publication process. I should also like to thank the Razumkov Centre's intern, Stefan Sokolowski, whose invaluable help in preparing the text is greatly appreciated.

NOTICES

*****

This manuscript was funded by the U.S. Army War College External Research Associates Program. Information on this program is available on our website, *http://www.carlisle.army.mil/ssi/*, at the Publishing button.

*****

*****

Comments pertaining to this report are invited and should be forwarded to: Director, Strategic Studies Institute, U.S. Army War College, 122 Forbes Ave, Carlisle, PA 17013-5244. Copies of this report may be obtained from the Publications Office by calling (717) 245-4133, FAX (717) 245-3820, or by e-mail at *SSI_Publishing@carlisle.army.mil*

*****

All Strategic Studies Institute (SSI) monographs are available on the SSI Homepage for electronic dissemination. SSI's Homepage address is: *http://www.carlisle.army.mil/ssi/*

*****

ISBN 1-58487-170-9

CONTENTS

FOREWORD

Ukraine has been a pivotal actor in Eurasia since its independence in 1991. Ukraine's destiny is critical to the security of the entire post-Soviet zone, and both it and the United States always have acted on that assumption. The stated goal of Ukrainian defense policy long has been to integrate with Euro-Atlantic structures like the North Atlantic Treaty Organization (NATO), and this goal has been one of the chief objectives of the United States, as well. However, to move from rhetoric to implementation is particularly difficult where the defense reform of a post-Soviet state has been concerned, and Ukraine is no exception. Nevertheless, in the past few years Ukraine has begun to make remarkable progress towards its self-professed goal of defense integration with Western structures.

Mr. Leonid Polyakov's detailed study of Ukrainian-U.S. defense relations and of Ukraine's defense reforms provides a comprehensive account of these two intertwined processes with focus on the last 5 years. His analysis clearly points out both the obstacles and the successes that Ukraine has encountered in its defense reform and outlines the challenges ahead for both partners. Given that Ukraine is a major contributor to the stabilization forces in Iraq and a key player in any European and Eurasian security order, this monograph is of more than academic interest. It has great policy relevance, especially as the United States seeks to work with its allies and partners in other post-Soviet states to foster their defense and political integration with the West. This monograph deserves careful consideration, and the Strategic Studies Institute offers it to foster better understanding of Ukraine's pivotal role.

DOUGLAS C. LOVELACE, JR.
Director
Strategic Studies Institute

INTRODUCTION

From the earliest times of its post-Soviet independence, Ukraine has been open to security cooperation with the United States. In the beginning, there were significant differences in political, security and even bureaucratic cultures between the two countries, which formed some obstacles to building bridges quickly. Many of these obstacles remain, especially in the political dimension of relations between the two countries. But in the absence of their former ideological differences and united by common interests in preserving international peace and fighting terrorism, Ukraine and the United States have established constructive and mutually beneficial military cooperation.

The United States has been interested in engaging post-Soviet Ukraine in security cooperation and clearly articulated what it wanted to achieve from this cooperation. It was in U.S. interests to have a strong, independent, stable, and democratic Ukraine as a partner in Eastern Europe. Guided by such a vision, the United States consistently has demonstrated initiative in supporting Ukraine in building its national military by engaging it in peacetime military-to-military contacts. The Ukrainian government unhesitatingly accepted U.S. leadership in bilateral military cooperation, which has provided it with an opportunity to learn useful approaches to defense reform, raised Ukraine's international prestige, and strengthened the country's position vis-à-vis the pressure for regional influence exerted by its neighbor (and regional dominant power), Russia.

Bilateral programs of military contacts with the United States have become the largest among Ukraine's many international military cooperation programs. Since 1992 bilateral military cooperation has improved in terms of quality and substance, and set the stage for preparation, execution, and support of actual U.S.-Ukraine combined operations in Bosnia, Kosovo, and now Iraq. These combined deployments have demonstrated that the years of cooperation were not in vain; Ukrainians have proven their ability to be a reliable and capable peacekeeping combat force.

However, as this monograph suggests, despite steady improvement in bilateral cooperation, developing full interoperability between the Ukrainian and U.S. militaries beyond joint peacekeeping is not yet a realistic possibility. At a time when full combat interoperability is beyond reach for even the closest U.S. allies, the experience of previous U.S.-Ukraine partnership shows that the most logical and realistic option is to promote and further improve tactical interoperability for low intensity conflict: peacekeeping, peace-building, and humanitarian assistance. More ambitious goals are far beyond Ukraine's current financial capabilities, and are restrained by the country's inability to qualify politically and economically for North Atlantic Treaty Organization (NATO) membership soon.

This monograph consists of four parts. Part I presents a strategic context for U.S.-Ukrainian military cooperation and provides general data on the history and current state of security relations between Ukraine and the United States. Part II focuses on the development and improvement of cooperative mechanisms for bilateral military contacts. Part III provides data and analysis of peacetime military engagement and discusses important lessons learned. Part IV examines Ukraine's practical cooperation with the U.S. military in operations in Kosovo and Iraq—operations where cooperation continues today.

In sum, U.S.-Ukrainian military cooperation has created a reasonable foundation for limited joint and combined action, with the **United States helping Ukraine to build a noticeable cooperative capability**. This capability currently is being adjusted in Iraq and other places. The potential remains for even greater cooperation, if necessary improvements are made.

The United States should not be expected to carry the burden of the future international peace and security agenda alone. In exercising its leadership, the United States will have to rely on ad hoc coalitions as often as it will rely on its closest allies. **Ukrainian troops, though not among the closest U.S. allies, are a likely partner of the U.S. military in future contingencies**. Thus the success of U.S. future engagements could depend on how the two countries act today to build their interoperability.

The history and lessons of U.S.-Ukrainian military cooperation may be of interest to scholars in post-Cold War East European security affairs, and to operational planners and practitioners who are creating and/or participating in a coalition force including the United States, Ukraine, and/or other post-Soviet or post-totalitarian states.

PART I

STRATEGIC CONTEXT

> We are like long-distance cousins; where you grew up and you really did not know your cousin, and now for some reason you have both moved to the same city, and now you have the opportunity to see one another. And now, as you get to know each other, certain things are very important and demand attention right now, and then later on, there are still items to work on, you just have to spend the time to get to know each other better. And with time, eventually, you will fall in love with each other. So maybe it is not long-distance cousins. Maybe it is long-distance lovers.
>
> Joel Ostrom
> Interview, November 12, 2003

In terms of territory, Ukraine is the largest European nation. It is strategically located at a European crossroads between the eastern borders of enlarged NATO and the European Union (EU) and the western border of Russia, with Belarus to the north and the Black Sea to the south. Ukraine's border with Russia is 2,063 kilometers (km) long; with Belarus, 975 km; with Poland, 542.5 km; with Slovakia, 98 km; with Hungary, 135 km; with Romania, 608 km; and with Moldova, 1,194 km.

Ukraine's population is approximately 48 million—down by 4 million from what it was in 1991, when it became independent after the dissolution of the former Soviet Union.

Ukraine is famous for its fertile black soils and, consequently, for its agricultural sector ("bread basket"); Ukraine produces millions of tons of such commodities as grain, sugar, sunflower oil, and meat. Ukraine is also blessed with a wealth of natural resources such as iron ore, manganese, coal, nickel, and uranium. Except for oil and gas, Ukraine is basically self-sufficient. It is both agricultural and industrial, producing a wide range of products including spacecraft, tanks, radars, ships, transport aircraft, and many other state-of-the-art products. Its workforce generally has earned a good reputation in Europe for being industrious, educated, and skilled.

Ukraine also could become a notable sea power. Among the Black Sea littoral nations, Ukraine holds first place in terms of its coastline's length—1,720 km (Russia, 920 km; Turkey, 1,320 km; Georgia, 290 km; Bulgaria, 260 km; Romania, 240 km). The country has 18 seaports and 8 shipyards, as well as its own system of educating maritime specialists.[1]

Politically, Ukraine is a unitary state and presidential-parliamentary republic. According to its constitution, Ukraine is a democratic, social, and law-based state. All the formal signs of a democratic state exist: political parties, an elected Parliament, and an elected president. But in reality, Ukraine is still struggling to become a real democracy. Contemporary political life too often becomes a battleground between dominant oligarchic and opposing leftist and national-democratic forces, with major democratic values very often declared rather than enforced.

Yet, as a forward step, Ukraine announced its intention to integrate into NATO and the European Union in May 2002.

Historical Background.

Given its crossroads position between Europe and Asia, and its natural wealth, it is no wonder that this land historically was a battleground for legions of conquerors and for its natives, who lived through centuries of glorious victories, humiliating defeats, heights of national spirit, and darkness of subjugation and oppression.

On the territory currently occupied by Ukraine, the first Eastern Slavic State, Kievan Rus, emerged in the early 9th century, with its capital in Kiev. The state's first rulers traced their roots to the Varangians (Vikings), who came to Kiev from the north. The history of this medieval state is largely a consecutive series of war stories about fighting invaders and internal rivals, conquering neighboring tribes, and threatening Byzantium across the Black Sea. In the 10th century Kievan Rus formally adopted Eastern Orthodox Christianity as the state religion.

The state existed until the 13th century when it became a common heritage for Ukrainians and Russians. In 1240 Kiev was destroyed by the Mongols/Tartars, who dominated this land for about a century until they were pushed away by the Grand Principality of Lithuania, and later by the Rzech Pospolita (a Polish-Lithuanian medieval state). The northern territories of Kyivan Rus around Moscow were separated, and for almost another 200 years they were dominated by Tartars, later becoming the Principality of Moscow, and later still—"Great" Russia.

In the Ukrainian part of the Polish state, the introduction of serfdom in the 16th century led to the emergence of a phenomenon called Cossacks. These were former peasants who could not tolerate serfdom and chose to escape to the southeastern steppes on the lower River Dnipro (Dnepr), where they established a stronghold called the Zaporozhian Sich.

With time, the Roman Catholic Polish state accepted the existence of the Sich and often used its free peasant-soldiers in its wars. But later, the Orthodox Zaporozhian Sich gained strength as a political power and pursued an independent policy of shifting alliances with powers other than Poland. This became intolerable for the latter and provoked attempts at suppression. The subsequent Cossacks' rebellion in the middle of the 17th century under the leadership of Bohdan Khmelnytsky was victorious, but the Cossacks failed to secure an independent state, choosing instead in 1654 to join Moscow's Orthodox Tsar, on the condition that they preserve a confederate status and internal self-rule.

Moscow later reneged on this arrangement and imposed its power on Kiev, liquidating Ukraine's autonomy and proclaiming itself the ruler of Ukraine. In 1775 the Zaporozhian Sich was destroyed. "Muscovites" became "great Russians," or simply "Russians," and Ukrainians became "little Russians" and lived under Russian and later Soviet rule.

While relations between ordinary people of the two nations throughout 3 centuries have been generally tolerant, Russia's imperial rulers were always on alert for the stirrings of Ukrainian nationalism and tolerated no hints of Ukraine's escaping their political control. The Ukrainian language was once banned in the 19th century, and signs of national independence were always persecuted severely. In the 20th century's Soviet period, this took the form of the destruction of Ukraine's attempts to become independent during the civil war of 1917-20; the imposition of an artificial famine in 1932-33 after Ukrainians showed that they were unwilling to submit to collectivization (some seven million Ukrainians perished from hunger); and the severe persecution of Ukrainian nationalist resistance after World War II.[2]

In 1991, after the infamous coup in Moscow, Ukraine's Parliament (Verkhovna Rada) adopted an Act of Ukraine's independence on August 24, and a national referendum on independence in December showed that 91 percent of the population supported independence. Thus the modern state of Ukraine appeared on the world map, although the prospects for its independent development were not rosy.

Ukraine inherited from its Soviet past the aftermath of the 1986 Chernobyl nuclear disaster; a large, but "heavy" energy-inefficient industrial base, with some 80 percent of it dependent on Russian-made components and energy supplies; an almost total lack of national-level management; and a weak civil society. Ukraine also inherited Soviet bureaucratic traditions and an uneasy relationship with Russia, which, despite the official proclamations of friendship and strategic partnership, at times could be characterized as uneasy cooperation—a "cold peace."

Ukraine as a Military Power.

In terms of military power, newly independent Ukraine inherited a great deal from the Soviet Union.[3] Forces located on Ukrainian territory were part of the second strategic echelon of the Soviet western theatre of operation. In pure numbers, the country inherited the world's third largest nuclear arsenal; about 40 percent of the former Soviet Union's armed forces personnel and equipment; and huge stocks of Soviet strategic reserves of arms, supplies, and ammunition.

As for ground forces, in 1991 Ukraine hosted on its terrain 5 armies, 1 army corps, 18 divisions (12 motorized rifle, 4 tank, and 2 airborne), 3 airborne brigades, and 3 artillery divisions. The Air Force had four air armies, while Air Defense had one air defense army and three corps. Also, a Black Sea Fleet, not initially under Ukraine's control, was divided between Ukraine and Russia in 1997. There was one rocket army of nuclear-capable Strategic Missile Forces and many support units. Soviet military educational facilities on Ukrainian territory consisted of 34 military colleges and 78 military faculties at civilian universities to provide military education and training.

In terms of numbers, there were 6,500 battle tanks; more than 7,000 armored combat vehicles; 1,500 combat aircraft; 270 attack helicopters; 350 combat ships and support vessels; and millions of small arms pieces and millions of tons of ammunition. Ukraine inherited a large nuclear arsenal, with 220 strategic weapon carriers, including 176 land-based intercontinental ballistic missiles (ICBMs) (130 SS-19 and 46 SS-24 missiles) and 44 strategic bombers (19 Tu-160 and 25 Tu-95). The overall nuclear potential of the strategic force was estimated at 1,944 strategic nuclear warheads, in addition to 2,500 tactical nuclear weapons. The ICBMs were targeted at the United States and armed with multiple independent reentry vehicle (MIRV) warheads, and every bomber carried long-range cruise missiles. In the military were some 800,000 troops in total.[4] Ukraine also possessed one-third of the Soviet defense industry—1,840 enterprises and research centers that employed 2.7 million people. Some facilities had unique capabilities such as shipbuilding and missile production.

However, it was not a coherent national defense sector per se, but a fragmented collection of what used to be the mammoth Soviet military-industrial complex, which had to be transformed into the national defense sector of Ukraine. Despite Ukraine's becoming the world's third largest armed power immediately after gaining independence (taking into account all inherited conventional and nuclear capabilities), it did not have much time to enjoy this status. The extremely high "inheritance tax" on this Cold War legacy became obvious very soon. In fact, the legacy brought more debts to be paid than wealth for prosperity.

It was obvious from the beginning that Ukraine needed a smaller force for its defense and one that it could afford. But in the absence of sound experience and appropriate intellectual capacity to produce a meaningful reform plan, Ukraine initially tried to recreate a Soviet-model system. These efforts, however, took place against a background of rapid economic decline. Due to poor governance and the virtual absence of effective economic reforms, the promising economic potential of Ukraine rapidly disappeared: from 1991-95, Gross Domestic Product (GDP) dropped by almost 50 percent. It bottomed out in 1999 with an estimated 60 percent reduction, if compared with the last preindependence year of 1990. Thus in the 1990s defense reform basically was confined to cutbacks of the most obsolete and expensive parts of the Armed Forces (for numbers, see Table 1).

	1991	1996	2002	1991-2002 Total Reduction	2006 Plan
Battle tanks	6,500	4,080	2,467	3,033	1,800
Armored combat vehicles	7,000	5,050	4,103	2,897	3,500
Combat Ships	9	12	17	+8	18
Combat aircraft	1,648	1,090	392	1,236	225
Attack Helicopters	271	230	133	138	120
Total military personnel	800	400	290	510	160
(Army personnel), thousand	(300)	(190)	(170)	(130)	(106)

Table 1. Dynamics of Reduction of Weapons, Military Equipment, and Personnel in Ukraine.

By the turn of the millennium, however, the situation qualitatively changed. All nuclear weapons were removed or dismantled. Economic decline was also reversed in 2000, and since then Ukraine's economy has grown by some 5-8 percent per year. Necessary defense reform experience also was acquired and applied to reach more reasonable decisions. Though the military equipment remaining in service has become 13 years older (with poor maintenance and scarce modernization), more attention and resources are coming to some priority areas, providing hope for the positive outcome of reform efforts. Constant attempts to reform the military generally have resulted in certain structural changes and in the threefold reduction of the numerical strength of the military—from almost 800,000 military servicemen in 1991 to 250,000 in 2003. However, the country is unable to sustain even this number, and further reductions by almost half are imminent. On the positive side, a certain degree of experience has accumulated, and there have been modest successes in some rather low-tech/low-cost areas (international peacekeeping; combat engineering; chemical, biological, and radiological [CBR] protection; military education; and transport aviation).

In addition, recent encouraging decisions point Ukraine's military towards the creation of a Joint Rapid Reaction Force (RRF)[5] and to the transition to all-volunteer manning instead of the current mixed conscript/contract model. Defense budgeting has shown modest signs of improvement, which has made it possible to channel resources to funding some priority requirements, such as, for instance, military intelligence and RRF unit training.

To a significant extent, Ukraine's limited achievements in the defense sector are due to the country's active participation in international peacekeeping and military-to-military cooperation, both under the NATO Partnership for Peace (PfP) program and under programs of bilateral military cooperation. After the terrorist attacks on the United States on September 11, 2001, Ukraine became a member of the antiterrorist coalition.

Beginning with 400 peacekeepers sent to the former Yugoslavia in 1992, in 2004 Ukraine had well over 3,000 peacekeepers around the world participating in 11 missions in such places as Iraq, Lebanon, Liberia, Sierra Leone, Kosovo, Georgia, and others. Having gained a reputation as a reliable peacekeeping force, the military of Ukraine is in constant demand from the United Nations (UN) and other security organizations, thus significantly contributing to Ukraine's international image. Ukrainian transport aviation is working hard under contracts from the UN and from individual NATO countries (including the United States) to sustain operations in Africa, Afghanistan, Iraq, and other places.

With regard to cooperation with NATO, Ukraine became the first of the Commonwealth of Independent States (CIS) countries to join the NATO PfP Program in February 1994. In 1995, Ukraine-NATO cooperation started in a real peacekeeping operation, with the deployment to Bosnia of the Implementation Force (IFOR). It continues—beginning with the follow-on Stabilization Force (SFOR), and later with the Kosovo Force (KFOR), in which Ukraine has almost 250 personnel, mainly within a joint Ukrainian-Polish mechanized battalion.

On July 9, 1997, in Madrid, Spain, Ukraine and NATO signed the "Charter on a Distinctive Partnership between the North Atlantic Treaty Organization and Ukraine." This event paved the way for the creation of a Joint Working Group on Defense Reform and a Joint

Group on Civil Emergency Planning. Soon after signing the charter, NATO opened its Information and Documentation Center in Kiev (its first on the territory of the former Soviet Union), and established a NATO liaison office in Ukraine's Ministry of Defense, which has been headed by a retired U.S. military officer since January 2004.

In 2000 Ukraine ratified the PfP Status of Forces Agreement (SOFA) and in 2002, the Exchange of Classified Information Agreement, and offered its large Yavoriv Training Center in Western Ukraine for the purposes of PfP training and exercises. Ukraine's NATO partners render important assistance in personnel training and developing interoperability. Ukrainian representatives are regularly invited to attend NATO training courses abroad and participate in joint exercises. Many hundreds of Ukrainians already have graduated from colleges and courses abroad.

Since 1995 Ukraine has participated in the NATO interoperability-building vehicle commonly known as the Planning and Review Process (PARP). It started by accomplishing only one interoperability objective out of 14 the country voluntarily took upon itself to achieve in the first cycle of PARP (1995-97), and somewhat improved during the second cycle—11 accomplished out of 27 selected. Currently Ukraine is in its third PARP cycle program, and is trying to integrate PARP, previously limited to the interoperability of PfP-designated units only, into the overall process of reforming Ukraine's military. To this end, Ukraine has coordinated with NATO, through the Joint Working Group on Defense Reform, to develop a list of 51 PARP partnership goals, prioritizing language training, command and control, and logistics. The most illustrative practical examples of these efforts are the creation of a multinational staff officers course,[6] increased emphasis on English language proficiency for certain categories of officers' positions, and creation of the first all-volunteer units within RRF.

Overall, Ukraine's partnership with NATO has developed into an extensive, mutually beneficial, and promising cooperation. Taken together with such major factors as the historical military tradition of Ukrainians, the growing contribution of the Ukrainian military to international peacekeeping, and recent positive developments in Ukraine's defense reform, it creates a favorable background for building U.S.-Ukrainian military cooperation and the particular issue of improving interoperability.

Political-Military Framework.

It probably would not be an exaggeration to say that, before independence, Ukraine, for a majority of educated Americans, basically was known—if at all—only as the one-time "bread basket" of Europe and as a part of the former Soviet Union (Russia). This knowledge reflected only Ukraine's agricultural potential—a historical image created by largely Russophile-dominated American East European studies.

For a few, Ukrainians also were known for their original culture and many centuries of continuous but largely unsuccessful struggle for self-government and independence. In the course of this struggle, Ukrainians sometimes were very ruthless to their perceived oppressors—be it Poles, Russians, Germans, Jews, or Turks—and these stories helped to inflate a false, stereotyped image of Ukrainians as "stubborn and ungrateful" or "greedy and selfish" nationalists. In addition, as part of the Soviet Union/Russia, Ukraine was known as a territory harboring a large part of the Soviet nuclear potential, which was targeted at the United States, and as such, represented a vital threat to U.S. security (which was correct).

Given that Ukrainians themselves had a rather limited ability to present their historical arguments, their influence on U.S. foreign policy prior to becoming independent was also limited, being realized only through *émigré* circles. The above-mentioned dominant perceptions of Ukrainians as "unpredictable and dangerous nationalists" were very likely behind the logic of the famous speech in August 1991 by U.S. President George Bush to the Supreme Soviet (Parliament) of the Ukrainian Soviet Socialist Republic just weeks before the infamous Moscow coup and Ukraine's declaration of independence, in which he warned Ukrainians that ". . . freedom is not the same as independence . . . [Americans] will not aid those who promote a suicidal nationalism based on ethnic hatred."[7] Such warning naturally surprised Ukrainians and ignited protests and demonstrations from the Ukrainian Diaspora in the United States.

Thus the beginning of the political relationship between the two countries was rather rocky.[8] Despite Ukraine's earlier adoption of the Declaration of State Sovereignty in July 1990, which declared that "Ukraine would not acquire or produce nuclear weapons," it took some time for U.S. foreign policy towards Ukraine to change its main focus from pressuring Ukraine to denuclearize.

The difference in political, security, and even bureaucratic cultures was also a noticeable obstacle to building bridges fast enough,[9] but several major developments contributed to a positive change in the relationship, which generally took shape by 1995. First, Ukrainian authorities appeared to be cooperative on the issue of nuclear disarmament. They did not always seem eager to satisfy their U.S. partners, since they did not consider it appropriate for Ukraine to bear the entire burden of dismantling and disposing of nuclear arms. After all, it was not Kiev's decision to deploy these arms on Ukrainian territory. However, the successful conclusion of the denuclearization agreement with the United States and Russia (the so-called Trilateral Accord) and ratification of the Strategic Arms Reduction (START) I Treaty and the Lisbon Protocol in early 1994, soon followed by Ukraine's accession to the Non-Proliferation Treaty, effectively made the country a non-nuclear state and defused all nuclear-related concerns.

Second, Ukraine appeared to be open to security cooperation with the West, specifically with NATO and the United States. In fact, the exchange of visits, invitation of military students, and other cooperative events between Ukrainian and U.S. militaries started immediately after Ukraine became independent and well before the signing on July 27, 1993, of the first background document for military cooperation between the Ministry of Defense of Ukraine and the Department of Defense (DoD) of the United States—"Memorandum of Understanding on Defense and Military Contacts."

Third, despite all the political and economic turmoil of the first years of independence, Ukraine managed to avoid the internal discord and even bloodshed common to many other CIS states such as Azerbaijan, Georgia, Moldova, and Russia. Ukraine also set the first and very important real example for the rest of the CIS countries by managing the peaceful transition of presidential power in July

1994 from Leonid Kravchuk to Leonid Kuchma.

The new, much more positive spirit of cooperation was reflected in high-level U.S. policy documents. If one looks into the U.S. *National Security Strategy* for 1995 (subtitled "A National Security Strategy of Engagement and Enlargement," February 1995), one cannot help but notice how often the state of Ukraine is mentioned in this 27-page document—over 20 times. The Strategy also introduced a policy of distinguishing Ukraine, in parallel to Russia, from other newly independent states when talking in general terms about the countries of the former Soviet Union.

Thus an initial period of suspicion and cautious rapprochement lasted for the first half of the 1990s but paved the way to a following period of much broader political and security engagement and cooperation, which continued through the second half of the decade. During that period the United States and Ukraine achieved significant progress in the political and security dimensions of their bilateral relationship. The main institutional structure overseeing this progress was the U.S.-Ukraine Binational Commission created in September 1996 and led by President Kuchma and Vice President Gore. In the corresponding declaration on the establishment of this Commission, the term "strategic partnership" was used for the first time.

In fact, throughout the 1990s the United States became the major global lobbyist for Ukraine.[10] In 1996-97 Ukraine even became the third largest recipient of U.S. assistance in the world (after Israel and Egypt), and the number one recipient in the former Soviet Union. But this did not continue for long due to the U.S. Congress' disappointment in Ukraine's failure to resolve specific disputes involving U.S. firms, and the country's general inability to provide favorable conditions for U.S. business development and investment.

By the year 2000, what had become the ritual practice of mentioning of Ukraine in parallel to Russia in official speeches and communiqué had faded away. There appeared to be a sense of growing disappointment by Washington in Ukraine's overall progress. In the U.S. view of that time, Ukraine:

> . . . has not moved from independence and nuclear disarmament to economic and political reform . . . has not escaped the predation and

> corruption of the transfer of power and property from state to private hands . . . its leaders have spoken confidently of a European future for Ukraine, but they have done little to prevent Ukraine from slipping into a near Third World country, which for the old, weak and marginal is a decline from the Soviet past.[11]

The unfolding Ukrainian political crisis was precipitated by the murder of opposition journalist Georgiy Gongadze in the fall of 2000. Seeing no progress in the resolution of the case and a deepening political crisis for more than a year, by 2002, America's new administration began to show signs of frustration with and loss of interest in Ukraine. The U.S.-Ukraine Binational Commission ceased to function, and the previous cordiality of political relations was clearly evaporating.

Other problems followed. The United States had misgivings over Ukrainian supplies of heavy armaments in 2001 to the Former Yugoslav Republic of Macedonia during the internal ethnic strife there. (The weapons sales were legal and above-board, but the United States viewed them as potentially destabilizing). But the real "bombshell" was the announced authentication by U.S. authorities in 2002 of a recording in which President Kuchma allegedly approved the proposed transfer of the Ukrainian "Kolchuga" air defense system to Iraq. This created a full-fledged political scandal. The U.S. Government responded by delaying the appropriation of Freedom Support Act funds, while NATO allies downgraded Ukraine's participation in the NATO-Ukraine Commission meeting at the November 2002 NATO Summit in Prague from the head of state to ministerial level. In the latest *National Security Strategy* of the United States (September 2002) prepared by the Bush administration, Ukraine is not mentioned at all, in noticeable contrast to the *National Security Strategies* of the Clinton era. Ukrainian authorities, for their part, denied all weapons sales accusations and, in turn, showed displeasure towards Western philanthropic organizations allegedly funding the Ukrainian political opposition.

Fortunately for the two countries' partnership, converging interests prevailed over these differences. By the start of 2003, political decisions on both sides were made that these setbacks should not prevent a mutually beneficial U.S.-Ukrainian relationship.[12] Ukraine agreed to deploy its CBR-protection unit to Kuwait in March-

April 2003 during the campaign in Iraq, and later contributed a peacekeeping brigade to the stabilization efforts of the coalition force there. This was not easy for Ukraine for a number of reasons, mainly due to negative public attitudes towards the war in Iraq and the strong opposition of Russia.

The U.S. leadership responded by agreeing to stop ignoring Ukraine's leadership and to meet briefly with Ukrainian President Kuchma, whose international image suffered seriously as a result of Gongadze- and "Kolchuga"- related scandals, and with Ukrainian Prime Minister Yanukovych during visits to the United States in 2003. U.S. officials also expressed cautious support for Ukraine's hopes of becoming a NATO Membership Action Plan (MAP) participant in the near future. However, they made clear that the success of future bilateral relations would be judged by the progress of Ukraine's democratic development.[13]

Thus, in the middle of the third 5-year period of their relationship, in the political dimension the United States and Ukraine appeared to be adopting the policy of small practical steps towards each other, rather than attempting to revive the inflated expectations of "strategic partnership." In light of their recent experience, both sides are very careful to keep the door of cooperation open in spite of possible disagreements.

Security Cooperation.

While political relations between the two countries have had a great many high and low points, sometimes resembling a foreign policy "roller coaster," their security cooperation somehow managed to remain rather stable and has been progressively improving. This cooperation has developed along three main axes: dismantling Ukraine's Soviet nuclear legacy, improving the national security of Ukraine, and cooperation in fighting terrorism and peacekeeping.

As far as cooperation on dismantling of the Soviet nuclear legacy is concerned, it started immediately after Ukraine became independent. Even the issue of Ukraine's diplomatic recognition was conditioned by U.S. concerns about nuclear weapons on Ukrainian soil, which were addressed during the meeting between U.S. Secretary of State

James Baker and Ukrainian President Leonid Kravchuk on December 18, 1991, prior to President Bush's announcement on December 25, 1991, of the formal U.S. diplomatic recognition of an independent Ukraine.

It took 3 years (1992-94) of relentless negotiations for the United States and Ukraine to bring the latter firmly to non-nuclear status and frame the generally successful nuclear disarmament process. In June 1996 the last nuclear warhead left Ukrainian soil, in October 2001 the last Ukrainian ICBM silo was destroyed, and the same year the last of the Ukrainian Tu-160 *Blackjack* and Tu-95 *Bear* strategic bombers were destroyed, along with hundreds of air launched cruise missiles (ALCMs) or converted to nonmilitary use. In August 2002 the 43rd Rocket Army of the former Soviet Strategic Missile Forces was finally disbanded. Thirty-eight nuclear capable Tu-22M *Backfire* bombers were also destroyed, while the remaining 19 Tu-22Ms were scheduled to be eliminated by September 2004. The most visible remaining issue is the U.S. possible support for the destruction of rocket engines, which is anticipated to be concluded in 2007.

Overall, by the start of 2004 the United States will have spent almost $700 million to eliminate strategic nuclear delivery systems in Ukraine within the framework of the well-known Cooperative Threat Reduction (CTR) Program (also known as the "Nunn-Lugar" program after its U.S. congressional sponsors). According to former Ambassador of the United States to Ukraine Carlos Pascual, "It is almost easier to predict where this country will be in 25 years than in 3 years." That, he added, is precisely why the United States must keep supporting Ukraine's efforts to eliminate its weapons of mass destruction. "It is the best security money we have ever spent."[14]

U.S. support for Ukraine in improving its national security is another dimension of cooperation that is 13 years old and continues to develop progressively. Its major component is defense cooperation, but also includes export control and nonproliferation and other areas such as border control, emergency management, and civil-military cooperation.

In the second half of the 1990s, bilateral programs of military cooperation grew to more than 120 events every year, which made the U.S.-Ukraine programs of that time almost twice as large as Ukraine's second largest bilateral program (with the United

Kingdom) and almost three times larger than Russia's program of military cooperation with Ukraine.

A large proportion of the bilateral cooperative events were funded within the scope of the CTR Program as well—through targeted CTR monies promoting "Defense and Military Contacts Programs" focused on promoting defense reform and developing trust and interoperability. However, after September 11, 2001, the number of yearly events funded this way within Ukraine's armed forces has been cut approximately in half. This reduction reportedly occurred "due to postponements caused by the new priorities under the Global War on Terrorism and deliberate efforts by the U.S. and Ukraine sides to focus on quality rather than quantity."[15]

But this reduction did not occur in other funding sources and programs, some of which even increased and thus partly compensated for the reduction in the major traditional source. For instance, in 2002, each U.S. regional Combatant Command (U.S. European Command [EUCOM], for example) has a source of U.S. DoD funding known as "Traditional CINC (commander-in-chief) Activity"[16] (TCA) funds, often utilized for cooperative activity with partner military forces. TCA, for example, funded two exchange visits between the U.S. Army Europe's (USAEUR) Southern European Task Force Brigade and the Ukrainian 95th Separate Airmobile Brigade, the main ground forces component of Ukraine's rapid reaction force. These exchanges established the groundwork for a partnership between the brigades to be further developed through the 2003 PfP exercise "Peace Shield."[17]

The U.S. International Military Education and Training (IMET) Program has provided over U.S.$16 million since its establishment in Ukraine in 1992 to educate about 500 Ukrainian officers and noncommissioned officers (NCOs) and over 100 civilians in U.S. military establishments (See Table 2). IMET training supported efforts to improve interoperability between the Ukrainian and U.S. and NATO militaries, and to promote transformation and restructuring within the Ukrainian Armed Forces, by providing opportunities for select Ukrainian officials to attend U.S. military and educational institutions. In addition, the expanded IMET (E-IMET) Program provided training for Ukrainian military and civilian officials, including personnel from nondefense ministries and the legislative branch, on defense budget management, creating an effective military justice system and moving to civilian control of the military.

Fiscal Year	Training Expenses $Thousand	Number of Students Total	In that, Civilians
1992	0,075	3	0
1993	0,413	13	0
1994	0,600	8	1
1995	0,707	40	23
1996	1,020	33	9
1997	1,015	37	9
1998	1,250	46	16
1999	1,250	48	4
2000	1,338	43	2
2001	1,443	49	5
2002	1,675	53	10
2003	1,692	48	13
2004	1,700	77	19
2005	1,800	TBD	TBD
Total (by end of 2004)	15,978	498	111

Table 2. The U.S. IMET Program for Ukraine: General Indicators.[18]

In the Ukrainian case, the IMET Program has achieved qualified success. It might be judged as less successful if compared with participants from Ukraine's neighbors in Central Europe, but it is certainly more successful if compared with Russia. Several Ukrainian graduates of the IMET program have achieved prominent official positions such as Minister of Foreign Affairs Kostiantyn Hryshchenko (Naval Post Graduate School), former Head of the Analytical Service of the National Security and Defense Council of Ukraine Anatoliy Grytsenko (Air War College), and a few ambassadors, deputy ministers, and parliamentarians. Others have attained the rank of general officer such as Sergiy Mokrenets (Army War College), Oleg Taran (National War College), Leonid Holopatiuk (Naval Post Graduate School), and several more.

Ukraine's interoperability with U.S./NATO in peacekeeping and humanitarian operations, (for instance, Ukrainian involvement in KFOR) has been supported through programs such as the U.S. Department of State—Foreign Military Financing (FMF) Program and U.S. Department of State—Peacekeeping Operations. In particular, this support is made available to Ukrainian participation in the Ukrainian-Polish Battalion (UKRPOLBAT) peacekeeping support operations within the U.S. sector of Kosovo. This funding is not only helping to solve NATO manpower needs, but provides Ukrainian troops experience in interoperability with Western forces and exposure to NATO professional standards and practices.

Every year some $1-3 million is allocated by the U.S. Departments of State and Defense to Ukraine's participation in PfP Program through the so-called "Warsaw Initiative," founded by the United States in July 1994. Among other important purposes, the initiative helps to promote the ability of the Ukraine's armed forces to cooperate with NATO allies and NATO partners, as well as to prepare for joining NATO.

In terms of security cooperation, the United States also has provided significant technical support in nonproliferation efforts and strengthening Ukraine's export control system. It supported the strengthening of Ukraine's borders against illegal migration and cross-border crime, as well as in many other projects. Recently, the United States approved a $30 million multiyear comprehensive CTR weapons of mass destruction (WMD) Proliferation Prevention Initiative of equipment, training, and technical assistance for the State Border Guard Service and the State Customs Service. Rather than individual parts, the Border Guard Service was asked to develop a complete concept including land, maritime, and riverine control. Once the concept is developed, the equipment needed will be provided as an integrated solution beginning in 2004.

Through the California-Ukraine partnership, training is underway at the California Highway Patrol and Border Guard Academy at the city of Khmelnytskyi. Other exchanges and expert visits are planned for 2004 to address contraband cargo inspection measures and rapid response unit operating models. An interagency team of California law enforcement agencies support these and other needs, including criminal intelligence sharing tools from the Western States Information Network, officer safety from the California Department of Justice, and harbor security from the San Francisco Police Department.

This tremendous amount of U.S. security support to Ukraine paid off on a number of occasions. This included Ukraine's agreement in 1998 to terminate its participation in the Russian-Iranian contract on the construction of a nuclear plant in Bushehr, Iran; Ukraine's contribution to the antiterrorist coalition through granting overflight rights and sharing intelligence information; and, of course, the contribution of peacekeepers to Bosnia, Kosovo, and Iraq.

These examples demonstrate that the 13 years of Ukraine's bilateral military cooperation with the United States and other NATO countries were not in vain; they created at least a minimal foundation for combined action. The Ukraine-U.S. and Ukraine-NATO programs proved their value, which could have been greater, if not for certain political and administrative problems on the Ukrainian side. Prominent East European security affairs expert British scholar James Sherr, speaking about Iraq, noted, "Like the United States, Ukraine has shown nerve. Very few others have shown it. If it can keep its nerve and defend its interests on other fronts, it will not only earn the support of the 'new Europe', it will earn its right to be part of it."[19]

PART II

THE COOPERATION MECHANISM

> To categorize it as "good or bad" overall, however, I would have to say "good," but I am disturbed at aspects of the "negative side" of the equation. There exists still a tendency in the Ukrainian Defense bureaucracy to over-centralize planning and decisionmaking. In any system this inevitably leads to various inefficiencies and frustrations.
>
> Harry Simmeth
> Interview, December 11, 2003

From the beginning, the initiative in bilateral military contacts was clearly on the U.S. side. Several obvious factors prompted this situation, such as the much heavier political and military "weight" of the United States vis-à-vis Ukraine, more substantial U.S. experience in building cooperative relations with other countries, a U.S. commitment to finance the bulk of expenditures related to cooperative events, and so on.

Most importantly, the U.S. side had a conscious interest in engaging Ukraine and a rather clear understanding of what it wanted to achieve from this cooperation. It also had systemic strategic guidance documents that framed "engagement" with emerging states as a key policy and that provided the "glue" to create and hold together an appropriate cooperation mechanism. The U.S. *National Military Strategy*, in particular, fleshed out the basis for military-to-military engagement programs in support of the national strategy.

On its part, Ukraine simply accepted U.S. leadership in planning and coordination of the bilateral military cooperation, although without any correspondingly well-coordinated strategy documents or guidance. It seemed to be enough that Ukraine was now an independent democratic state and no longer had any ideological differences with the United States. Moreover, a comfortable relationship with the U.S. military represented at least a symbolic counterweight to Russian influence or bullying. On a practical level, military cooperation with the United States would provide an opportunity to learn useful approaches to defense reform, which for many also meant the very attractive opportunity to travel abroad, including to the United States.

In fact, the Ukrainians were cooperative from the very beginning and still are. But due to the continuous instability inside the Ukrainian military, precipitated by lack of clear political guidance as well as by permanent reorganizations and reductions, the military was very slow in building effective military cooperation structures and procedures. These structures and procedures still require significant improvement.

Strategic Guidance.

When the issue of U.S.-Ukrainian military cooperation first appeared on the agenda in 1992, the U.S. side already had a hierarchy of strategic guidance documents and certain institutional instruments to translate the guidance into mutually-negotiated plans for real cooperative events such as the IMET and FMF programs, as well as military-to-military contact visits and even combined exercises. At that time, the global strategic equation had just changed from bipolar confrontation to an uncertain and ambiguous global environment. In the wake of the end of the Cold War and disintegration of the Soviet Union, the immediate U.S. security interests in Eastern Europe were very clear: reduce the threat of nuclear war, constrain the proliferation of WMD, preclude any reemergence of the "evil empire," and support regional stability. The subsequent shift from previous strategies of containment of communism and nuclear deterrence naturally required a new policy of flexible engagement.

Thus the existing U.S. institutional and intellectual capability to formulate and articulate U.S. national interests helped to start the cooperation with Ukraine and other so-called newly independent states of the former Communist bloc. It also augmented the mechanism of foreign military relations with several high profile overarching security programs such as CTR and PfP, as well as additional targeted military-to-military cooperation programs, including the U.S. European Command Joint Contact Team Program (JCTP)[20] launched in 1992 (formally it got to Ukraine in 1998) and Army National Guard State Partnership Program (SPP)[21] which started in 1993.

However, when Ukraine became independent, it had no mechanisms for strategic planning and planning to protect national interests by military means. Ukraine had no National Security Council, no Ministry of Defense, and no General Staff of the Armed Forces. Capabilities of the previously existing Ministry of Foreign Affairs of Ukraine were rather limited, given its small staff, republican subordination within the former Soviet Union, and the routine practice of the outflow of the best specialists to Moscow.

The first Ukrainian strategic level documents were soon drafted and approved by the Parliament, but their value was very limited because of their largely declarative nature. This was further exacerbated by the complexities of the nation-building processes in the 1990s, which took the form of long internal political battles for redistribution of powers between Ukrainian legislature and the executive branch.

Ukrainian national security strategy level documents like "The Foundations of Ukraine's Foreign Policy" (1993) and "Concept of the National Security of Ukraine" (1997) were rather good as the first national exercise in drafting strategic level conceptual documents. However, the correct, albeit "very general in tone," provisions reflected "the continuing ambiguities present in defining Ukraine's security interests, threats, and policy objectives."[22] More to the point, the virtual absence of an enforceable implementation system to force executive structures to follow these documents, plus continuous internal political rivalry in the country, precipitated the big gap between the documents' theses and real practices in security area.[23]

At least these documents defined the strategic basics to include the non-nuclear, nonaligned status of the country and its general interest in European integration and building cooperative relations with the United States. But they certainly were too unspecific to provide clear strategic guidance for Ukrainian planners on policy priorities and allocation of resources.

As far as the U.S. *National Military Strategy* is concerned, in terms of its continuity from the U.S. *National Security Strategy*, the first post-Cold War document of 1995 displayed a clear example of how to translate the *National Security Strategy* requirement of "engagement and enlargement" into specific subordinate military strategy guidance:

> Military-to-military contact programs are . . . effective instruments . . . to forge new and more cooperative relationships both with former adversaries and with formerly non-aligned nations. . . . success . . . hinges on mutual trust, effective communications . . . interoperability, and doctrinal familiarity . . . The militaries of Central and Eastern Europe are a particular priority.[24]

In comparison, the main Ukrainian military security level documents of the 1990s—"Military Doctrine of Ukraine" (1993) and "State Program for the Building and Development of the Ukrainian Armed Forces for the Year 2005" (1996)—provided less clear guidance, especially on the goals of military cooperation with any other countries. They generally reflected the euphoria and inexperience of the first years of independence.

Instead of providing specific direction, the 1993 military doctrine only contained general tasks and a long list of good intentions, with priorities unsupported by resources. It stated very ambiguously that Ukraine "builds its relations with other countries regardless of their social-political system and military-political orientation on the basis of recognition of all issues of national security of two sides."

The State Program (1996) was developed by the military itself, approved by the president, and was classified "secret." Its main tenets, which were made public, did not go very far. They included adoption of specific legislative acts, preparation for all-around unilateral defense of the country, and set general goals of force modernization and numerical reductions to the level of Conventional Armed Forces in Europe (CFE) Treaty limitations.

However, the turn of the decade brought new tendencies on both sides. In the United States, the new Republican Bush administration came to office. It suggested a new strategic vision, which, while generally continuing the "engagement and enlargement" approach, appeared to be shifting from inclusive multilateralism to "coalitions of the willing." Of course, the entire focus of U.S. strategy was affected tremendously by the events of September 11, 2001, and by the U.S. response. While reemphasizing common values, the *National Security Strategy* (2002) also puts special attention on accountability of the receivers of foreign aid: "Nations that seek international aid must govern themselves wisely, so the aid is well spent. For freedom to strive, accountability must be expected and required."[25]

In Ukraine, the experience of reforming the military, based in no small part on the useful examples of other countries including the United States plus changes in the global and regional security environment, brought about new national strategic guidance as well as a new national military strategy. The new "State Program for Reform and Development of Armed Forces of Ukraine," adopted in 2000, made significant steps in defining the priorities of defense reform to include creation of a Joint Rapid Reaction Force, gradual abolition of conscription and a turn to manning on a contract basis, and improving the Reserve structure.

Furthermore, in May 2002 the National Security and Defense Council of Ukraine took a strategic decision to join NATO in the future and in July 2002 the Ukrainian President approved the Strategy of Integration to NATO. After this decision was made, Ukraine developed and agreed on an Individual Action Plan with NATO. Ukraine also tried (thus far, unsuccessfully) to be admitted to NATO MAP participation.

Given the negative influence of the remaining problems of the country's democratic and economic development on the progress of defense transformation, these latest positive developments in Ukraine's strategic security guidance should not be overestimated. However, they are leading in the right direction and provide more clear guidance for military cooperation.

Planning for Military Cooperation.

In 1992 the United States began military cooperation by forming contacts with high-level Ukrainian military officials. Invitations to Ukrainian officers to study at U.S. military colleges under the long-existing IMET program followed. In 1992 the first two Ukrainian colonels went to the U.S. Army War College at Carlisle Barracks, Pennsylvania, and the U.S. Army Command and General Staff College at Fort Leavenworth, Kansas. Issues of future military cooperation were discussed when then-Ukrainian Minister of Defense General Kostiantyn Morozov visited Washington, DC, in April 1992, at the invitation of then-Defense Secretary Richard Cheney.

On both sides, specific structures overseeing new military cooperation issues were created. The United States assigned this task to the newly created "Russia, Ukraine, and Eurasia (RUE)" Division in the J-5 Directorate of the Joint Staff. This newly formed staff division consisted of five officers—about half the size of the typical J-5 division. The Ukrainians created the Foreign Relations Directorate (FRD) within their General Staff of the Armed Forces, where the mission of coordinating cooperation with the United States (and Canada) was assigned to a corresponding group of five people.

The bulk of the first contacts in the early years of military cooperation was related to the issue of Ukrainian nuclear arsenal dismantlement and primarily coordinated through Ukraine's Ministry of Foreign Affairs. However, the first high-level political and military contacts also secured the first bilateral military cooperation agreements, which enabled the planning and conduct of the first specific military-to-military cooperative events between the Ukrainian Ministry of Defense and the U.S. DoD. A Memorandum of Understanding (1993) provided for semiannual meetings of a "Bilateral Working Group on Defense and Military Contacts" and annual talks between the U.S. Joint Staff and Ukraine's General Staff.

During the period of 1993-95, the initial yearly plans of military contacts were worked primarily through countries' defense attaché offices, and the number of bilateral cooperative events slowly grew from some 20 in 1993 to around 50 in 1995. In fact, despite the high interest on both sides in large expansion of military cooperation, the agenda of Ukraine's nuclear disarmament naturally overshadowed the agenda of bilateral military-to-military contacts. This issue remained until the United States and Ukraine worked out all the major problems of Ukraine's nuclear weapons by the end of 1994. Then, in spring 1995, Defense Secretary William Perry visited Ukraine and suggested moving past issues of nuclear disarmament to questions of social protection, security, and military and economic cooperation.

Colonel Harry Simmeth,[26] the U.S. officer who became responsible for the execution of this policy guidance, found the task rather difficult, given that he reported to the Chairman of the Joint Chiefs of Staff (CJCS). Under U.S. law the CJCS does not command military assets; primarily he is the chief military advisor to the President and the Secretary of Defense. Normally, one of the regional U.S. Combatant

Commands would assume the job of military cooperation with a country in its area of operations. In Ukraine's case, this role is held by the USEUCOM. However, during the Cold War, the entire Soviet Union (including Ukraine) was never assigned to a U.S. Combatant Command, a situation that continued until 1998. This meant that CJCS had at his immediate disposal few resources with which to execute military contact programs, and could not simply order the regional Combatant Commands to assist. (Another effect was that, while a Combatant Command would normally have a sizeable staff section coordinating military cooperation, CJCS basically only had Colonel Simmeth. It also meant that some of the most mundane and routine matters, rather than being handled by "experts" at lower levels, immediately rose to the attention of the highest levels.

Essentially, CJCS—meaning, in this case, Colonel Simmeth—had to "ask" the Combatant Commands and other U.S. agencies to support proposed programs and events with the necessary resources. This vastly complicated his ability to construct and execute a plan. He was largely reduced to proposing events to these agencies and soliciting their ideas. His one advantage was the availability of the additional funding sources noted earlier, which at least heightened the "spirit of cooperation" shown by the commands and agencies solicited. What almost necessarily emerged initially from this process was largely a "buffet" of proposed activities, rather than a set of coordinated activities within a coherent plan. Colonel Simmeth observed:

> There were no formal lower-level planning mechanisms beyond the BWG (bilateral working groups) and Joint Staff Talks through which to develop coherent programs. But, since there were those mechanisms at the strategic level (BWG and Joint Staff Talks), the higher leadership had agreed in general to conduct activities. In turn, there was pressure to elaborate this into actual cooperative events. This created pressure for results. Planners were expected to devise various events whether there was a coherent plan or not. Worse (from a planning perspective) the pressure for such tangible "deliverables" was not tied to any rational planning schedule, but primarily to the date of the next high-level meeting.[27]

However, despite the fact that the Ukrainian side participated enthusiastically in developing this first formal agreement on military cooperation (negotiated by Colonel Simmeth and a Ukrainian counterpart in 1995 for implementation in 1996), it was by far primarily the U.S. side that suggested the majority of cooperative events contained in the document. Although it was referred to as the "U.S.-Ukraine Plan for Military Cooperation for 1996," it was (as noted) more properly a list of events based on the guidance available in U.S. national security documents, concepts, and ideas generated by J-5 RUE's counterpart staff section in the U.S. Office of the Secretary of Defense (OSD), CJCS' intent, and the "volunteered" services of U.S. various commands and agencies.

This said, it nevertheless must be noted that the "plan" was far from being totally disjointed, because of relatively clear guidance provided by U.S. strategic level documentation. This guidance specified the goals of "encouraging military reform," "civilian control of the military," "fostering greater regional cooperation," and "facilitating Ukraine's participation in PfP." It was assumed in this documentation that military contact events could support these goals by focusing on their substance while developing "mutual trust, effective communications, interoperability, and familiarity with each other's doctrines." Another goal, "imparting democratic values," in the process would happen primarily by familiarizing Ukraine's military with the democratic processes in which the U.S. military is required to work, and by personal interaction among commanders, staffs, and troops during cooperative activity.

Additionally, there was some further rationalization of the Plan during Joint Staff-General Staff negotiations. For example, the effort was made to link certain events (as when a visit to the United States by an Ukrainian IL-76 *Candid* transport plane for training with the U.S. Air Force brought along a Ukrainian Army Regimental staff for an exercise with the U.S. Army). An attempt also was made to identify activities that suggested follow-on events for the next year's list, in the hope of creating logical progression (for instance, the exercise "Peace Shield" 96 would be a U.S.-Ukraine sponsored event geared to help prepare Ukraine to host a larger and more complicated NATO exercise in following years).

When the plan was developed, the Ukrainian side agreed with all of the U.S. proposals; this surprised the United States, which doubted Ukraine would be able to handle such an ambitious schedule of activity (this turned out to be true, with about 15 percent of the events being cancelled or postponed). But the larger surprise—and eventually, disappointment—was the realization by the United States that the Ukrainian side was not particularly strong in defining its own goals for bilateral cooperation with the United States, even when asked and "coached."

Given the Ukrainian lack of clarity regarding priorities, the United States mounted a much more serious effort over the next several years to help them focus on meeting both U.S. and Ukrainian objectives and producing a "programmatic" approach, resulting in a more integrated and coherent cooperation plan. This led to the creation of a working-level direct contact arrangement among military planning experts called the "Colonel's Conference." This conference was co-chaired by representatives of the U.S. Joint Staff and the Ukrainian General Staff, and comprised delegates from each side's military services, commands, and appropriate agencies, to include a prominent role for USEUCOM. This model helped to produce contact plans that were much better focused, and, despite the general expansion of the plans, to reduce pressure for "numerical success." The list of events was now backed by the first coherent sub-programs, such as plans for the continued long-term conduct of the "Peace Shield" series of peacekeeping exercises, the "Sea Breeze" naval exercise series, and a highly comprehensive project to help create a Ukrainian NCO corps. California and Kansas were designated as PfP state partners for Ukraine; their state National Guard organizations cooperated in creating a series of coordinated events focused on civil-military affairs, disaster relief, and search and rescue, as well as peacekeeping and contingency operations. In 1998, an important change took place in the U.S. command and planning arrangements—Ukraine finally became assigned to USEUCOM (along with Ukraine's neighbors, Belarus and Moldova, plus the Caucasus region). This had a significant impact in that both sides had to adapt to new cooperative mechanisms. The Colonels' Conference was replaced by the Military Liaison Team (MLT)—an arrangement successfully tested by USEUCOM in Central Europe.[28] The idea was for a team of U.S. officers to work in Kiev with Ukrainian officers, as well as with the USEUCOM Staff,

> to create plans, identify resources available to fulfill activities, and to submit timely requests to other agencies. Regularly established channels would be available for efficient funds administration. From the U.S. perspective, this would be a much more efficient, continuous, and coherent way to operate.[29]

In general, however, because of the U.S. friendly but somewhat dominating attitude to Ukraine in the 1990s, plus Ukraine's lack of clear focus on what it wanted from U.S.-Ukrainian military-to-military cooperation, the initial plans developed under the MLT concept concentrated on familiarization events to the detriment of events tied to specific goals and objectives. To some extent, the MLT program seems to have arrived with the notion that the parties were "starting anew" despite the record of cooperation already logged. On a larger scale, however, the NCO development project continued to be successful, and there was a further rationalization of exercise goals and schedules.

To be completely fair to the U.S. MLT concept, the cooperative attitudes of the Ukrainians were dealt a blow by the reorganizing of FRD into the Department of International Cooperation of Ukraine's Ministry of Defense (DICMOD). DICMOD replaced the FRD as the agency interfacing with the MLT. The U.S. concept had been to align the MLT with the Ministry of Defense/General Staff for direct liaison, the model used in East Central Europe. Ukraine instead imposed the DICMOD, which initially seemed more concerned with building its own "empire." This included stressing its own "importance" by demanding that all contact, even routine correspondence, be routed through its own very inefficient bureaucracy. Inevitable delays and frustrations developed, as did the quality of planning overall. In an environment of generous U.S. funding, coupled with Ukraine's inability to prioritize and (eventually) DICMOD's bumbling, bilateral military cooperation frequently witnessed inflated expectations on the Ukrainian side and produced frustration for the United States. Even strong U.S. supporters of the military cooperation program were often at least irritated by what often appeared to be the Ukrainians' overestimation of costs and efforts to "squeeze" one project to fund additional activity. (For example, to accommodate a U.S. ship visit, a request was made that the United States provide funds to refurbish the pier at which the ship was to dock). Former Army attaché in Kiev

Colonel Timothy Shea has skeptically concluded about that period of cooperation that:

> The amount of money thrown at peacetime military engagement has convinced senior Ukrainian leadership that the United States has unlimited resources and that the decreasing incentives represent Washington's indifference. What is needed is less lecturing, greater humility, more thoughtful organizing, rewarding positive change, and discouraging inappropriate action.[30]

A legacy of Soviet-style behavior to this day remains among many Ukrainian top military leaders and some remaining Soviet KGB-style security regulations, as well as a legacy of Soviet centralizing bureaucracy. This has furthered U.S. frustration. As Colonel Shea observed,

> Designed to be collocated with counterparts on the general staff, MLT in Ukraine's case was forced to accept residency on the opposite side of Kiev from Ministry of Defense. Instead of directly coordinating with planners, the team relies on DICMOD *apparatchiks* to administer the program . . . The legacy of the Soviet armed forces and KGB remains deeply imbedded in the psyche of most senior officers.[31]

To make the situation worse, the unfolding severe economic crisis in Ukraine in the 1990s made it impossible to achieve any Ukrainian defense reform objective beyond reduction of structures, personnel, and equipment. The humiliating social conditions of Ukrainian officers provoked a noticeable outflow of qualified personnel from the military in general, and in particular from structures coordinating military contacts, where personnel had comparatively high marketing value due to the knowledge of foreign languages and possession of valuable experiences. There was also a problem of so-called "military tourism"—distorted criteria for selection of participants in cooperative events would frequently occur, such as when an appropriate "expert" would be passed over in favor of someone in a position to "bump" him. This was especially true in case of trips from Ukraine to other countries, which at that time also meant at least per diem pay—an important addition to a Ukrainian officer's monthly pay of lower than U.S.$100, even for colonels.

However, the turn of the decade brought better strategic guidance on the Ukrainian side and a slightly different, more pragmatic strategic focus on the U.S. side, which, along with experience gained, helped to improve the planning and management of cooperative events on both the U.S. and Ukrainian sides.

To cure the problem of exaggerated cost estimates, for example, the United States, since FY 2000, moved from direct reimbursement of goods and services provided by the Ukrainian side during combined exercise, to hiring contractors and providing contracts based on competitive bidding.

To provide for better coordination on its side, the United States brought together the previously separate MLT (responsible for JCTP and SPP) and the Security Assistance Office (responsible for IMET and FMF), and in 2001 created in Ukraine a joint structure—the Office of Defense Cooperation (ODC), which also incorporated the Humanitarian and Civic Assistance Office.

To provide for a better quality rather than a greater quantity of cooperative events, the number of planned events was reduced twice from over 120 at the end of the 1990s to some 60-70 "high quality" events per year starting in 2001. Plans themselves were drafted with more discretion and have been more closely tied to defense reform priorities under a new Ukrainian State Program (2000).[32] New issues of cooperation of a rather technical nature appeared on the rise since Ukrainian participation in the Kosovo (from July 1999) and in the Iraq (from August 2003) campaigns, for which the United States provided partial financial compensation and technical assistance.

To facilitate the Ukrainian military's force development and acquisition prioritization planning, in 2001 during Defense Secretary Donald Rumsfeld's visit to Kiev, the United States suggested conducting a defense analysis of the Ukrainian Armed Forces by a team of USEUCOM experts. The focus was Ukraine's RRF. The analysis, accomplished by October 2002, provided valuable background both for Ukraine's defense reform planning and for planning of bilateral military contacts. According to the Head of Bilateral Affairs Office (the new name for MLT after being moved under ODC) Major Joel Ostrom,

> For 2004, our planned focus is on the Rapid Reaction Force that was agreed to by the U.S. and Ukrainian leadership at the last joint staff talks.

> The reason for that is to ensure that the effort, the financial resources, are going to be used to further professionalism of the Ukrainian military, which will last beyond its restructuring and the downsizing of the Ukrainian military.[33]

On the Ukrainian side, qualitatively positive actions could be found in development of the system of military contacts coordinating structures and somewhat better personnel management. In addition to the central bilateral cooperation body of Ukraine's MOD (now the DICMOD), each service now has its own division of military cooperation. In the General Staff a separate Directorate of Euro-Atlantic integration was created with the specific mission to coordinate Ukraine's military NATO- and EU-related activities, as well as to participate in peacekeeping missions. A top official at the level of Deputy Defense Minister was designated to address foreign military cooperation issues.

The detrimental Ukrainian tendency to attempt to obtain as much money from the United States as possible has been tempered. It is true that the United States is more attuned to the issues and provides better oversight. But the Ukrainian military seems to have somewhat changed its attitude as to the "unlimited" nature of U.S. resources. During the conference in Warsaw in May 2003 to discuss contributions to the Iraqi Stabilization Force, Ukrainian generals announced the intent to contribute a brigade, but afterwards they were less eloquent (in notable contrast to some other contributing nations) in asking for either assistance or reimbursement of the brigade's deployment costs.

Unfortunately, other damaging factors on Ukrainian side remain, such as the tendency towards over-centralized decisionmaking, inefficient and irritating bureaucracy, and over-restrictive security regulations. But at least these problems have been identified, and discussion of them gains momentum; and with it, the chances for correction grow.

However, in this case, it should be pointed out that success will be difficult to achieve, if efforts to repair the culture of bureaucracy and secretiveness is limited to the Ukrainian military establishment only. To a significant extent, success of the military's democratizing remains hostage to success in wider democratic governance reform in the country.

In general though, despite all the remaining problems, the planning mechanism is improved, and now the two sides' cooperation encompasses not only peacetime military engagement, but preparation, execution, and support of actual combined deployments and operations. It is unlikely that the U.S.-Ukraine combined operations in Bosnia, Kosovo, and now Iraq would have been possible without the stage having been set by years of peacetime military engagement, whatever its imperfections.

PART III

PRACTICAL PEACETIME ENGAGEMENT

> There are a few aspects of interoperability that we must address when we discuss the bilateral U.S.-Ukrainian relationship. Technical issues can be addressed by making a commitment to procure interoperable systems. But physically obtaining western equipment does not solve the problem. Training and accepting doctrine are as, if not more, important. There needs to be a focus on western standards. This is where there seems to be the biggest gap.
>
> John Cappello
> Interview, December 30, 2003

Ideally, in order to be successful, practical military cooperation should be well-planned, well-supported, and well-executed in accordance with sound strategic guidance. According to *U.S. Joint Vision 2020* (*JV2020*):

> Although we must retain the capability to act unilaterally, we prefer to act in concert with our friends and allies. Laying a solid foundation for interoperability with our alliance and potential coalition partners is fundamental to effective combined operations. We remain committed to doctrinal and technological development with our key allies and to combined training events and exercises that contribute to interoperability.

Apparently the guiding principles of *JV2020* for jointness ultimately will guide the interoperability agenda as well:

> A fully joint force requires joint operational concepts, doctrine, tactics, techniques, and procedures—as well as institutional, organizational, intellectual, and system interoperability—so that all U.S. forces and systems operate coherently at the strategic, operational, or tactical levels. Joint effectiveness does not mean that individual pieces of equipment or systems are identical, but rather that commanders are not constrained by technical or doctrinal barriers among the components of the joint force, and that the joint force's capability is dramatically enhanced by the blending of complementary Service capabilities.

As far as any Ukrainian documents' provisions on interoperability are concerned, this issue is related almost exclusively to the Ukraine-NATO agenda. For instance, the *NATO-Ukraine Action Plan* (2002) states among the "Principles of Defense and Security Sector Reform" that:

> Cooperation in defense-related areas promotes interoperability with NATO and increases Ukraine's overall ability to be a key player in regional security. Reform efforts and military cooperation also support Ukraine's strategic goal of Euro-Atlantic integration by gradually adopting NATO standards and practices, and enhancing interoperability between the armed forces of Ukraine and NATO forces, in particular through the implementation of Partnership Goals and participation in NATO-led crisis response operations.

To this end, several objectives are formulated, such as:

> Develop the full interoperability, sustainability and mission effectiveness of the Armed Forces through effective implementation of Partnership Goals; maintain the readiness of Rapid Reaction Force units for participation in joint operations with NATO, and training of these units to meet NATO standards; achieve a required level of compatibility for the actual and future armaments and military equipment and doctrine of the Armed Forces of Ukraine, which allows minimum interoperability in order to conduct, on a case-by-case basis, tasks of common interest with NATO, and adapt/adjust acquisition and related practices to those of NATO Allies; and develop interoperability between Ukraine and NATO communication and information systems.

Provided that the two sides have enough trust in each other to contemplate fighting side by side, success in achieving interoperability between their militaries in an actual mission will be enhanced through practical peacetime cooperation. The actual level of interoperability thus becomes a derived category of the level of achieved common standards in language, training, doctrine, and technical systems during their practical peacetime engagement.[34]

Language.

From the very beginning, there was a clear understanding on both sides that truly interoperable allies should use the same working language and the same terminology; in other words, that use of English is critical to enhance interoperability. (Although both English and French are the official operating languages of NATO, English predominates by far. As it is unlikely that NATO will adopt Ukrainian as an official language any time soon, the importance of English, therefore, is obvious).

When the first Ukrainian IMET students went to the United States in 1992, it was already clear that an in-country system of preliminary English language training was needed by Ukraine. The intention was to train Ukrainian instructors at the U.S. Defense Language Institute and supply language laboratories to Ukraine, where trained instructors could bring Ukrainian IMET candidates and other related personnel to required proficiency levels before they went abroad for further study.

In 1993 the first two instructors went to the Defense Language Institute (Lackland Air Force Base, San Antonio, Texas). Ten years later, there were 32 U.S.-trained English language instructors, and several dozens trained in Canada and the UK. A number of foreign instructors—5 from the United States and 14 from the UK—came to the country through this period to teach English for Ukraine's military.[35]

During the first years of cooperation, the focus was on training candidates for the IMET program. But Ukraine's participation in peacekeeping operations soon broadened this focus. By 1995 two U.S.-supplied language laboratories were operational in the Odessa Army Institute and the Kiev Military Institute of National Shevchenko University, while the Kiev Army Institute offered 20 hours of language training for officers selected for peacekeeping operations. In 1997 one more U.S.-supplied laboratory was opened in the Sevastopol Naval Institute to support the exclusive needs of Ukraine's Navy. Since 1994 Canada and the UK have invited some 30-40 Ukrainian officers and MOD civilian employees every year to their language training centers as well.

In addition to IMET and other foreign programs, in the new context of Ukraine's commitments to an interoperability agenda under NATO's PARP, the Ukrainian military decided gradually to increase efforts to produce a sufficient number of qualified English-speaking personnel.

While more in-country language laboratories were needed and the United States planned to provide these, there was also an understanding that military colleges (in Ukraine they are termed military institutes) could and should provide better background training for cadets, which was to be further upgraded at language laboratories and specialized language courses when required. In 1998, Ukraine's Minister of Defense ordered the number of language training hours for cadets increased by 100 hours, and compulsory examination at the end of the last year was introduced.

By 2001, seven U.S.-supplied language laboratories were operational in Ukraine's military, along with several courses offered by the UK (the latter trained both active and retired personnel). However, even this number was enough only for training IMET candidates, while increased Ukrainian commitments to peacekeeping operations still suffered from a lack of language proficient personnel.[36] According to former Deputy Army Commander Lieutenant General Victor Hudym:

> [In 2001] English language as a common language for Army officers so far failed to become the focus of officers' education and training. Today we feel great demand for English speaking officers, especially for missions abroad and participation in PfP events. Unfortunately, in regular units the conditions for and organization of studying English cannot permit quality and timely accomplishment of joint missions and resolution of common problems.[37]

Indeed, as many Ukrainian and foreign experts observed, for Ukrainian staff officers it appeared much easier either to enroll in the IMET program or to attend language courses usually located in large urban areas, where the major Ukrainian military headquarters are located as well. But for units' commanding and logistical staff at bases in more remote areas, especially in the case of those having a large number of subordinate personnel requiring constant attention, it was nearly impossible to find the opportunity for language training.

Further, by the middle of 2002, after Ukraine declared its intention to become a NATO member, attention to English language interoperability assumed greater criticality. New actions followed. A system of language curriculums at educational institutions was supplemented by study courses streamlined into an "evening" (without interruption of service assignment), "day," and "foreign" system of courses.

Also in 2002, by the order of the Minister of Defense, a list of specific positions requiring knowledge of foreign languages was approved. This list contains over 2,500 positions. Provisions to make it easier for these selected personnel to attend courses also were developed and established. New incentives for those successfully mastering language were instituted to retain qualified personnel, since once an officer learned to speak English, his skill becomes highly marketable. However, "real life" conditions of unit officer assignment obligations often continued to be a stronger priority than declared requirements and incentives for language training.

The latest move came at the end of 2003, when the Minister of Defense approved the "Program for the strengthening of the language training of the Armed Forces of Ukraine personnel for 2004-2005." Under the program, the network of foreign-language courses is to provide places for 205 students in 2004. It will function based at 10 military higher education institutions of the Ministry of Defense and several selected training centers.

By the end of 2005, the network capacity is to be expanded and new foreign-language courses are to be created, including at locations where Ukrainian peacekeeping contingents are deployed. The overall network of language training is planned to have places for up to 320 students, thanks to these measures. Eight additional language laboratories, provided by the United States under the FMF Program will be established in military education institutions. There are also plans to introduce new ways of teaching classes without removing personnel from service. As a new development, not only officers will undergo language training, but other personnel also will participate, including selected soldiers, sergeants, and civilians.

In general, as of 2004, the highest and rather acceptable level of English language proficiency is being displayed by Ukrainian staff officers, many of whom accumulated experience through IMET education, local and foreign language courses, multinational exercises, and peacekeeping missions. Commanding and logistical staff officers coming from units generally have uneven levels of English due to time constraints not favorable to combining education with performing assigned unit-level duties. The poorest state of language proficiency is on the soldier-NCO level, where training is not organized properly as yet.

After 13 years of cooperation, the Ukrainian Armed Forces have made noteworthy progress in English language proficiency. Primarily, this is due to the continuous efforts on both sides, and a significant amount of U.S. financial and technical support. However, because of the lack of serious planning attention during the first years of cooperation and a lack of focus until very recently, language interoperability still requires significant improvement.

Training.

Since 1995, when the first joint tactical peacekeeping exercise, "Peace Shield," was conducted at Ukraine's Yavoriv Training Center, Ukrainian and U.S. Army units have been regularly involved in various kinds of training exercises. As a rule, these exercises are conducted "In the spirit of PfP," when Ukrainians and Americans play the key role in planning, financing, and conducting exercises, while inviting many other NATO partner nations to participate.[38] Not only U.S. and Ukrainian Army units have trained together, but also naval forces and marines/naval infantry ("Sea Breeze, "Eloquent Nugget"). Ukrainian and U.S. Air Force units additionally have conducted some low-level but productive combined training, especially in cross-familiarization with each other's transport aircraft and procedures. In addition, there is a number of now-traditional series of multinational exercises under the NATO PfP program, where Ukrainians cooperate with U.S. and other NATO militaries, contributing to U.S.-Ukrainian interoperability on different subjects to include maritime logistic support systems ("Cooperative Support"), communication system interoperability ("Combined Endeavor"), interoperability in peacekeeping and humanitarian relief tactics ("Cooperative Osprey" and "Cooperative Nugget"), and many others.

In relation to training exercises, it is important to note the unique role of the "Peace Shield" series, which became traditional. The overall number of participants ranges from 800 to 1,200, representing the armed forces of 20-26 countries. The cost of the exercise so far has been covered largely by the United States and constitutes some U.S.$400,000-700,000 per exercise.

With time, the scenario and equipment underwent changes and improvements. In 2000, the establishment of a satellite system of teleconference communication made it possible to keep certain participating units' headquarters outside the borders of Ukraine and locate them in Estonia and Bulgaria, from which locations they participated on a "virtual" level. In 2001, a transatlantic flight by three U.S. Air Force C-17 aircraft after three refuelings in midair ended with a joint Ukrainian-U.S. parachute drop on the Yavoriv Training Center by 180 paratroopers of the U.S. 82nd Airborne Division, 120 paratroopers of the 1st Ukrainian Airmobile Division, and 9 pieces of Ukrainian combat airborne fighting vehicles. The same year the exercise scenario changed from a traditionally abstract one to a concrete situation modeled on Kosovo and set in the zone of responsibility of the Multinational Brigade "East" under U.S. command, in which Ukraine participates.[39] The latter was truly an example of the progression to utilizing exercises to anticipate and solve practical issues of interoperability.

"Peace Shield" also was raised from the battalion to the brigade level and transformed from a live-training tactical-level event to both a live-training and virtual-training tactical/operational level event. This involved the introduction of computer-assisted staff-level exercise simulations with corresponding involvement of a small number of troops from the participating armed forces in live tactical play. Despite some continuing organizational deficiencies, these exercises provided valuable lessons learned for both the U.S. and Ukrainian sides. As noted in 1998 by U.S. Major General (Ret.) Nicholas Krawciw in after-action comments:

> For the future, it would be beneficial for NATO and Ukraine to conduct a series of computer assisted staff exercises designed to involve, over time, as many Ukrainian military headquarters as possible in various peacekeeping staff functions. Exercises "Peace Shield 97" and "98" provide a U.S.-Ukrainian model for useful future computer assisted training.[40]

The experience received during "Peace Shield" and NATO PfP exercises along with practical lessons from peacekeeping missions has allowed Ukrainians to produce several rudimentary tactical manuals for training their personnel to perform future peacekeeping together with other countries.

It also allowed, with support from the United States and NATO, the conduct of large-scale exercises like "Cooperative Adventure Exchange" in October 2002, when some 3,000 military personnel from 12 NATO nations (Belgium, the Czech Republic, Denmark, Germany, Hungary, Italy, Luxembourg, Norway, Poland, Turkey, the United Kingdom, and the United States) and from six partner nations (Azerbaijan, Bulgaria, Georgia, Moldova, Macedonia, and Ukraine) struggled to improve the interoperability of Allied Command Europe Mobile Forces, LAND with other headquarters and units from NATO and partner nations. At the same time these exercises promoted common understanding of deployment, organization, employment, and redeployment of a multinational formation in a fictitious UN-mandated NATO-led peace support operation.

The Ukrainians gained many good lessons from these combined exercises: peacekeeping techniques, standard operating procedures (SOP), the role and use of simulations, communication equipment, and leadership. At the same time, these exercises generally have shown that, while intellectually and physically Ukrainian servicemen were not much inferior to American, the Ukrainian combat training system, doctrine, and ethos of leadership needed significant improvement.

In fact, Ukrainian [post-Soviet] manuals provide a solid base of methods, procedures, and techniques for attaining necessary combat skills through rigorous training. This is one of the major reasons why the former Soviet military was considered to be very good in "generating raw combat power." As proven many times by the U.S. Army's Opposing Force (OPFOR) Regiment at the U.S. National Training Center (Fort Irwin, California), these techniques and skills, if mastered and applied properly, appear not at all to be inferior to those adopted in U.S. field manuals, and permit the OPFOR to fight and typically win tactical engagements with other U.S. Army units in training.

However, most people would agree that the average Ukrainian unit noticeably would be less combat capable than a U.S. one, even if given the same equipment. There is an obvious problem on the

Ukrainian side, which is a lack of necessary resources for training. In addition, conscript manning and less consistent leadership qualities could be counted as negative factors. But to a significant extent, the average Ukrainian unit would be weaker mainly because of the inferior training system currently in place. In essence, there is a big gap between what is written in manuals and what kind of real practice Ukraine inherited along with good combat manuals. On paper, units should master skills in accordance with the specifics of their assigned mission, geographical specifics of the territory where the mission is to be performed, and with different approaches to day/night conditions. In fact, this is hampered by a number of negative factors: significant amount of time consumed by repeating cycles of basic training for rotating conscripts; a very weak NCO corps and, consequently, Ukrainian officers being too busy with other than training business; little opportunity to train more than at company level—very basic but important things. And to top it all, Ukrainian units have no such things as developed Mission Essential Task Lists (METLs) to guide their training; in the U.S. military, each unit's METL helps to prioritize and assist in organizing training on the most essential tasks for that unit, which in turn helps to tailor and rationalize the current U.S. system of combat training.

In Ukraine, there is still a Soviet-style conscript-oriented system of "periods of training"—two half-year periods per year, based on the curriculum provided by "Program of Army combat training" and similar programs for other services. The system is driven in large part by the need to bring new conscripts "up to speed" every half-year. Thus, in reality, Ukrainian soldiers routinely master the same basic skills every period to pass the "control inspection" at the end, but very rarely is there an opportunity to be trained as a coherent unit in anything resembling advanced tactical operations. Ukrainian officers, faced with this simplistic non-innovative approach to combat training—especially when added to the very low quality of life provided by the current system of social support (pay, housing, retirement benefits, etc.)—often lose the incentive to try to train their units to any higher standards.

On a low level of joint peacekeeping, these differences are not a big problem yet, though it is a growing issue in terms of efficiency of the way Ukrainians man and train their higher-than-company peacekeeping contingents. This will be true especially in view of Ukrainian plans to have a fully professional RRF soon, as well a standing peacekeeping brigade of up to 3,000 strong and the deployment of a relatively large Ukrainian force in Iraq.

The necessity of improving the system of Ukrainian combat training was recognized by a U.S. team of experts, conducting the "Defense Analysis of the Armed Forces of Ukraine" in 2002. In the final report of this analysis, which was primarily focused on Ukraine's RRF, the team explicitly recommended to begin with developing METL and a new set of standards, and suggested using the U.S. Army set of manuals/evaluations "that may be useful as a model for a similar system for the Ukraine Armed Forces."[41]

As far as multinational headquarters staff training is concerned, the progress already is past the stage of familiarization. Growing participation of Ukrainian officers in peacekeeping deployments abroad has prompted the opening in 2000 of special courses for the officers of multinational staffs within the National Defense Academy of Ukraine. These courses train officers who are selected for the positions in international peacekeeping or similar multinational staffs. For each rotation, up to 40 officers study the theoretical foundations of peacekeeping for 2 weeks, as well as practical Ukrainian and international experience in working on a multinational basis. The curriculum is based on programs from NATO schools.

In 2003, the courses provided training for a group of Ukrainian officers who were to be appointed to various positions in Iraq, where 11 Ukrainian officers served in Joint Coalition Headquarters in Baghdad and 24 Ukrainian officers and one senior NCO in the multinational "Center-South" division headquarters in Babylon. Instruction was carried out exclusively in English, and the curriculum itself was adopted with consideration of actual operational experience in Iraq. Instructors from the United States, Canada, and France helped to teach the course, along with Ukrainian instructors.[42] There is a plan, with the support of NATO, to transform the courses for the officers of multinational staffs into a (international) center for training the officers for peacekeeping operations.

The United States supplied the courses with the bulk of supporting computers, local network equipment, software, and literature, as well as providing some visiting lecturers to supplement permanent NATO instructors from Canada and France. They also contributed to Ukraine's National Defense Academy program called "International Weeks," during which U.S. and NATO speakers lecture Ukrainian student-officers and their instructors on various security and defense topics.

In addition, on June 17, 2004, Ukraine Defense Minister Yevgen Marchuck and U.S. Ambassador John Herbst inaugurated the U.S. sponsored Simulation Center in the National Defense Academy, the first of several such centers supplied by the United States to Ukrainian military for the purpose of more efficient officer training for joint and peacekeeping operations. The overall cost of the U.S. supplied hardware, software, and training package is U.S.$5.4 million.

While Ukrainians go to the United States and some other NATO countries under training and exchange programs, Americans study in Ukraine, too. In 1997 two U.S. Army officers attended a 4-month course at Ukraine's Army Institute in Kiev (now in Odessa). This experience found its further application when U.S. officers and cadets from time to time have come to Ukraine to spend several months in a Ukrainian military unit or institute.[43]

Overall, in regard to training as part of the interoperability development process between the U.S. and Ukrainian militaries, past experience has proved that many problems of Ukraine's military training could be corrected. Decisions have been made to emphasize command and control interoperability: to introduce a Joint "J-structure" for Ukrainian headquarters and adopt NATO standards. Ukrainian officers, in fact, have two sets of standards—the old Soviet ones for combat and new NATO ones for peacekeeping.

In terms of the specific training of Ukrainian units, a broader use of training simulators is envisioned. But removing the negative impact on the training process of such factors as Ukrainian conscript manning, very limited resources, and virtual absence of professional NCO corps will require more time. If current positive trends continue, it will still take at least 5-10 years for the Ukrainian military to become qualified at a satisfactory NATO level.

Doctrine.

The problem of bringing Ukrainian military doctrines closer to the American ones was a need primarily emphasized by the U.S. side. At the turn of the decade, there was a feeling of some disappointment because of the alleged lack of enthusiasm on the Ukrainian side to utilize Western doctrine. With regard to the period of cooperation up to 2000, Colonel Simmeth summarized:

> Familiarity with each other's doctrines [is] partly successful. Ukraine's military is now more familiar with U.S. and Western doctrine through exercises, exchanges, and cooperation in the Balkans. The reverse is also true. But the hope in the West was that Ukraine would find Western doctrine useful in pursuit of military reform. It does not seem that this is the case.[44]

Colonel Simmeth's assessment was founded in the U.S. concept that doctrine tells a force what it needs to be able to do and to what standards. Thus from that, it may derive other conclusions such as what resource allocations, training programs, etc., rationally are necessary. In this sense, doctrine can serve as an "engine of change" for reform, particularly at the operational level and above. That Ukraine's limited reform progress up to 2000 was perceived at least in part to be hampered by a failure to reform doctrine thoroughly is not surprising.

Such a conclusion is natural as well for the United States in the area of interoperability, given its perception of the great importance of doctrinal compatibility for effective combined operations. U.S. Major General Krawciw noted, for example: "Doctrinal interoperability in joint and combined operations may well constitute the difference between ensuring the well-being and success of those sent into combat, or risking failure and loss of lives because of inadequate procedures and tactics."[45] Then Chairman of the Joint Chiefs of Staff General John Shalikashvili emphasized this idea to a visiting Ukrainian General Officer delegation in 1996 when the latter expressed the desire to obtain more U.S.-made military equipment "for the sake of interoperability." The Chairman explained his belief that we:

> do not all have to buy the same equipment to be interoperable . . . but our doctrine and techniques for employing that equipment should be compatible. This means having the same or similar operational doctrine, and understanding where that doctrine differs and why, as well as having compatible equipment standards.[46]

Overall, many Americans, after gaining experience in cooperating with Ukrainian troops, acknowledge their "fundamental" competence at individual and tactical levels plus good discipline. But at the same time, there is probably a consensus on the U.S. side about the weakness of Ukrainians in the areas of operational level doctrine and doubts about the operational competence of at least some of the mid-top leadership.

However, since the year 2000, and even during the prior period of bilateral military cooperation, the U.S. doctrinal influence on Ukrainian military was quite significant. Though Ukrainians always have attempted to base their system of developing major doctrinal documents primarily on their own intellectual resources, there was no doubt that experience received through military-to-military contacts played a very important role in this process. This experience came through many familiarization cooperative events, as well as through joint training exercises, the open sharing of U.S. doctrinal publications, and, of course, through graduates of the IMET Program.[47]

The U.S. experience undoubtedly has influenced the adoption of Ukrainian key doctrinal concepts like creation of the RRF and of the Active Reserve (State Program, 2000) and the 2002 decision on turning to all-volunteer manning by 2015. Occasional U.S. and other NATO countries'[48] doctrinal recommendations play an ever more influential role in many specific areas of Ukrainian defense reform.

For instance, EUCOM's Defense Analysis Report on the RRF has made a significant impact on the development of the Ukrainian manual, "Fundamentals of preparation and application of the Joint Rapid Reaction Force." Another good example, according to Ukraine's Minister of Defense Yevghen Marchuk, "The U.S. program DRMM (Defense Resource Management Model) is utilized in development of a well-balanced prospective force structure for the Armed Forces of Ukraine 2015 in terms of determining realistic resource allocations needed for this."[49]

Other individual NATO countries such as Canada, Germany, Poland, and the UK provide important support to Ukraine's doctrinal development in the form of general military reform analysis, training, and education of military personnel, personnel management systems, etc.

Starting from 2001, NATO teams of experts in cooperation with Ukrainians launched several pilot projects to develop recommendations for some specific areas like Ukraine's Naval force structure reform, reintegration of retired military personnel, and disposal of ammunition and small arms. In doctrinal terms, the most promising results and prospects are shown within a naval pilot project, where a NATO team in concert with U.S. experts has passed the stage of recommendation development and is helping Ukrainians to implement a new force structure in the Ukrainian Navy, establish a naval combat information center, adopt NATO standards for Naval headquarters, and improve the Navy's system of logistic support.[50] And, Western experience in general clearly was taken into account in the recently adopted new version of Ukraine's operational-strategic doctrine, "Fundamentals of preparation and application of the Armed Forces of Ukraine."

There are many recent indications of genuine progress by the Ukrainians on the operational doctrine level of interoperability. However, the issue of leadership remains for U.S. counterparts an area of some disappointment. The lower level leadership problem—reinforced by the weakness of the Ukrainian NCO Corps—was recognized by both sides long ago, and consequent priority attention was given to this, bringing visible improvements. The solution at higher levels is a bit more problematic.

Perhaps foremost, the decisionmaking style adopted by the senior Ukrainian leadership seems to frustrate the United States. It is a familiar thing to complain of the "Ukrainian tendency to over-centralize both decisionmaking and execution," which is in vivid contrast to the U.S. military doctrinal standard of "centralized planning and direction and decentralized execution." Some Americans recognize that this is not simply a case of bad leaders with poor leadership ability, but rather a cultural and organizational tendency, reinforced by a legacy of Soviet over-centralization. But it remains frustrating to most, who view it as an impediment to efficient planning and reform progress overall.

Another point of discontent, which has to do with adoption of the U.S. doctrinal experience, is the currently rather slow development of a Ukrainian system of "lessons learned." It took a long time for Ukrainians to recognize how important it is to have an effective multilevel system of lessons learned. While this problem finally was recognized, it is still unclear as to when Ukrainians will be able to fill the current gap between regular units' experiences and the ability of the central military research institutions to gather and distribute these experiences along with suggested "best practices." This problem is characteristic for both operational experiences and the way Ukrainians manage their resources. Prominent NATO experts on Ukraine Christopher Donnelly and James Greene have stated, with regard to the Ukrainian resource management problem which is of the same nature as "lessons learned" problem, that:

> More important than Rapid Reaction Forces or moving to contract service is the issue of managing resources . . . The system also needs to evaluate program results . . . For such system to work, it must be able to accurately measure and predict costs, both for current force structure and programs, and for alternatives. This is difficult in any armed forces; in a post-communist system the challenge is huge; intensive foreign assistance will be needed.[51]

Ukraine needs to adopt an effective system of reporting, analyzing, summarizing, and disseminating important lessons learned, as well as a methodology by which to compare these lessons to doctrine and thereby identify any doctrinal gaps or errors. U.S. experts suggest capitalizing much more on its experience in order to reduce waste of resources, prevent repeating the same mistakes, and disseminating innovative solutions and practices.

Overall, while Ukrainians actually adopted a lot of positive U.S. experiences for their Navy, military education system, and peacekeeping doctrine; the remaining problems with resources and bureaucratic shortsightedness noticeably slowed wider application of doctrinal lessons. It is important to remember, however, that there are natural limits to doctrinal aspects of interoperability, since in practical terms, it is still a bit early to talk about U.S.-Ukrainian military cooperation beyond peacekeeping, and Ukraine is not a member of NATO yet. According to many well-known expert estimates, even some U.S. NATO allies and other European partners often are considered not to be fully combat interoperable with the United States, and sometimes even not fully trusted.

In general, in terms of doctrinal interoperability, the two sides already are beyond the familiarization stage. Ukrainians started a number of important changes to adopt their force structure, staff procedures, peacekeeping operations, and logistical system to U.S. and NATO standards. However, problems remain with the full adoption of Western combat doctrines and with the leadership/decisionmaking style. When a specific joint mission is envisioned, Ukrainians simply adapt personnel to this mission's Rules of Engagement (ROE) and multinational staff procedures. The core of Ukrainian combat doctrine is considered to be not radically different or deficient to that of the United States. The comprehensive adaptation of manuals, headquarters' techniques, and leadership styles to U.S. and NATO models is not seriously on the agenda yet. And to ensure effective implementation of doctrinal change, there likely will need to be a longer-term deeper "cultural" change (and possibly a generational change) in the Ukrainian military and in the minds and styles of its members. Any more sudden radical "change of minds and styles" of the senior Ukrainian military leadership—especially when this is not strongly encouraged by their political masters—appears highly unlikely.

Technical Systems.

Given the scarcity of resources that the Ukrainian authorities are willing to allocate for the country's military armaments and equipment needs, the issue of technical systems interoperability has not been solved, not even by adoption of the "compatible equipment standards" noted by General Shalikashvili. The "solution" basically was left to the good will of the United States in donating or helping to fund the purchase of military equipment under its FMF program, the "Warsaw initiative," and other programs. In addition to Ukrainian financial limitations, it took a long time for the two countries to negotiate and sign the necessary background agreements on military-technical cooperation, which also blocked the development of more intensive contacts in this area. There is also a history of the United States buying small batches of Ukrainian weapons and other products for military use, but these amounts never exceeded several million dollars per year.

Still, in terms of cooperation in peacekeeping, the United States has provided quite a lot of equipment to make Ukrainian units technically interoperable. This concerns primarily Ukrainian military contingents in Kosovo and Iraq. For instance, under FMF, in FY 2000 Ukraine purchased HUMVEEs with tactical radios and other communication equipment. In FY 2001 Ukraine purchased additional HUMVEEs with tactical radios, other four-wheel-drive vehicles, and night-vision goggles.[52] In FY 2002 Ukraine used FMF funds to purchase, among other things, automation and simulation equipment and additional military tactical radios and communications systems, as well as to help develop a simulation center to support peacekeeper training.[53] This pattern continued in FY 2003. A certain number of radios and data-transmission equipment also were supplied to Ukrainian peacekeeping units from funds provided under the "Warsaw initiative."

This kind of support naturally helped to cover only the immediate needs of Ukrainian units to be interoperable with their U.S., Polish, and other partners in Kosovo and in Iraq. However, ambitions to join NATO pushed the Ukrainian military to declare its intention to equip all units with interoperable communication equipment, for which purpose the plans for international tender in 2004 were announced. Among the bidders Ukrainians expected to see well-established producers, such as Siemens, Motorola, Tadiran, and others.[54]

Ukrainians supplied some of their equipment to the United States as well, but these supplies were of a different nature. First, Ukrainians were selling, not donating, or helping anyone to fund their equipment. While the list of this equipment includes some Ukrainian tanks, combat vehicles, missiles, artillery systems, and even training aircraft, these sales have nothing to do with interoperability and were evidently purchased for different purposes. At a minimum these sales bring some initial experience of military-technical cooperation, which could possibly become helpful in the future, if

Ukraine continues to plan for acquisition of western interoperable equipment beyond communication systems.

Ukraine also provides the United States and other NATO nations with a significant amount of airlift, especially in cases of very large and very heavy cargo. It possesses a sizable fleet of Antonov-124 (*Ruslan*) and Iliushyn-76 *Candid* cargo aircraft, as well as the biggest plane in the world, the Antonov-225 (*Mriya*), capable of carrying a 240-tons maximum load. These aircraft were used in 2003 to deliver U.S. equipment to Kuwait and Iraq.

In fact, theoretical opportunities exist for future close military technical connections, given some of Ukraine's unique technological capabilities in the fields of space, transport aircraft, shipbuilding, missile, radar, and other production. But for this to become true, Ukraine will need to become a real "strategic partner." Beyond today's rather declarative political rhetoric, this is likely to occur only in the more distant future. This is the reason why, despite genuine interest on both sides to expedite military technical cooperation and despite available technological potential, the issue of technical systems interoperability probably lags behind most other main issues of interoperability.

Overall, the results of U.S.-Ukraine peacetime military engagement could be termed effective but not efficient. On the one hand, bilateral cooperative mechanisms have allowed for the conduct of a great number of useful familiarizing and training events and for the running several important joint projects. These events and projects have created the necessary conditions for the overall positive development of bilateral military cooperation and established a foundation for cooperation in practical deployments. On the other hand, so far the effect of these cooperative events on the progress of Ukraine's defense reforms has been mixed, and the process of strengthening Ukraine's military interoperability with U.S. and NATO militaries has proceeded at a rather slow pace. This reflects remaining shortcomings on the Ukrainian side in terms of improving language and doctrinal interoperability, as well as the rather insignificant efforts to improve technical systems interoperability. Most important, the Ukrainian side corrects problems only very slowly because its own internal system of the lessons learned still is very underdeveloped.

PART IV

ACCOMPLISHING MISSIONS TOGETHER

> Actual operational experience provides the greatest learning environment. Therefore, continuing Ukrainian participation in peace operations . . . should be considered as very important.
>
> Nicholas Krawciw[55]

Ukraine's participation in many UN and Organization for Security and Cooperation in Europe (OSCE) peacekeeping operations and in stabilization operations in Iraq makes it possible for the Ukrainian military to gain new experiences from working with the armed forces of other countries. Most importantly, this participation has tested the value of peacetime engagement and allowed Ukraine to improve interoperability with the United States and NATO partners, thus building the foundation for possible cooperation in future low intensity conflicts. The United States and Ukraine have cooperated most closely in accomplishing peacekeeping and combat missions together in Kosovo and Iraq. These experiences so far represent the most important instances of interoperability development, though many other operations could also suggest plenty of lessons from working together.

Kosovo.

At first, the Ukrainian KFOR contingent, consisting of the 14th Helicopter Company and the 37th Maintenance Company (replaced by a mechanized company in July 2000), was deployed to Macedonia, where it remained from July until December 1999. This time primarily was used to prepare for further deployment to Kosovo, to train troops for future missions, and to learn NATO standards and ROE. In December the Ukrainians moved to Kosovo, where they were based at Camp Bondsteell with the MNB "East" under U.S. command.

Among the many examples of joint actions between Ukraine and the United States in Kosovo, two are particularly interesting for study of interoperability. The first case involved the protection of the first democratic elections in Kosovo in October 2000. For KFOR's operational reserve, consisting of British and Greek infantry troops mounted on seven helicopters (three Ukrainian and four American), the mission was to move a multinational force anywhere throughout Kosovo quickly to provide extra security and additional forces for the municipal elections. To demonstrate the resolve of KFOR to secure peace and order prior to the election day of October 28, the strength demonstration, codenamed "Air Insertion Exercise," was conducted in an open field close to the town of Gnjilane.

The period of intense preparation was meant to demonstrate the readiness of KFOR's operational reserve to protect and secure the first free, democratic political elections in Kosovo. According to an observer's report:

> Coming in formations of three and three, the helicopters touched down in turn to drop off heavily equipped British and Greek infantry soldiers who ran bowed into their positions, waiting for the soldier deployment to be fulfilled. This joint helicopter force consisted of three American UH-60 *Black Hawks* and three Ukrainian Mi-8 *Hips*, all very well-coordinated and obviously familiar with this type of assignment. As soon as the last helicopter, an American CH-47 *Chinook*, touched the ground, the first formation of three helicopters took off again, immediately followed by the second formation. Shortly afterwards, the *Chinook* was also emptied and back in the air, chasing the six helicopters ahead. Left in the field were seven groups of soldiers.[56]

The scenario of the exercise was rather simple—it envisioned a demonstration of the unopposed deployment of the multinational KFOR Operational Reserve into a small field near the town. However, the message was powerful: "There should be no doubt that the Kosovo Force is ready to deal with any problems that might occur in connection with Kosovo's first municipal elections."[57] Judging by the conduct of the elections, it worked well.

According to Ukrainian peacekeeping experts, this episode in KFOR's mission marked a noticeable improvement in coordinated actions between Ukrainian and U.S. and NATO forces, compared to their previous deployments to Bosnia and Eastern Slavonia (Croatia) in the 1990s.

Another illustrative story which involved the Ukrainian and U.S. infantry took place in Kosovo on February 13, 2001. Ukrainian peacekeepers escorted a convoy of Serbian civilians through the mountains to the municipal center of Strpce. A sniper (allegedly Albanian) opened fire, apparently aiming at one of the bus drivers, but shot an elderly Serb instead. Ukrainian peacekeepers deployed immediately and searched the nearby slopes, where they found several Albanian males and detained them. Later in the evening a Serbian crowd of over 500 gathered around the local UN police station as a "spontaneous protest against the killing of the elderly Serb." The situation became tense as Ukrainian peacekeepers formed a circle around the station with the UN police contingent (including U.S. military police) inside while the crowd attempted to break in. At some point, the UN police station chief started firing in the air and aiming his gun at the crowd. This move ignited the mob—Molotov cocktails were thrown at the station, stones were hurled at Ukrainian peacekeepers, and five police cars were burned or destroyed.

The crowd was growing when Ukrainian company commander Captain Brezgounov entered the station with four soldiers and suggested the policemen evacuate. They hesitated when they learned that the Ukrainians were in small numbers. When the policemen were finally in Ukrainian vehicles, the crowd blocked their movement. Then support suddenly came from outside—U.S. HUMMVees rammed through the hastily-erected Serbian barricades on the outskirts of Strpce and raced to the aid of the Ukrainians. The senior U.S. officer present, Colonel Kamena, decided to address the crowd. As he approached, the crowd seized him, but Ukrainian peacekeepers led by the same Captain Brezgounov broke into the crowd and pulled the U.S. officer out of it. Within 24 hours, MNB "East" commander U.S. Brigadier General Quinlan awarded Captain Brezgounov with a NATO medal for outstanding performance.

This episode showed evident reluctance on the part of the UN police to cooperate with the Ukrainian KFOR troops. There was also some evidence of lack of coordination between the U.S. and Ukrainian troops once the action began. This could be attributed to a lack of proper advanced planning, but there was no doubt about the trust and support between the Americans and the Ukrainians.

Ukrainian veterans still gratefully recall that, immediately after this action, the U.S. troops supplied them with better antiriot equipment, which the Ukrainians initially lacked.

However, the two sides' military newspapers reported the incident quite differently. While the Ukrainian *Narodna Armiya*[58] did not hesitate to describe broadly and praise the role of both Ukraine and the United States, the U.S. *Stars and Stripes*,[59] in describing the incident, wrote a story about the role of the Americans, but said nothing about the role of Ukrainian peacekeepers that day, merely mentioning someone's proposal to assign U.S. troops as convoy escorts instead of unnamed "KFOR troops."

This difference in coverage might have to do with U.S. political leadership not wishing to publicize a danger to their troops, even though the U.S. forces on the scene recognized the action for what it was. This could be an example of trying to avoid the "*CNN* factor"—the story of U.S. soldiers being surrounded and having to be rescued by anyone could have produced public or political backlash. But this lack of publicity was perceived by some Ukrainians as reluctance to note their important role.

Since then, both the composition and the mission of Ukrainian contingent have changed. In 2001, Ukrainian helicopters were withdrawn, and Ukrainian troops were represented by two companies of the Ukrainian-Polish joint peacekeeping battalion, UKRPOLBAT, and the staff element. By the start of 2004, they are still serving within the U.S.-led MNB "East," patrolling the area near the Kosovo border with Macedonia along with the United States. The situation has become more peaceful, and the troops devote a great deal of time to training together. As observed by UKRPOLBAT Deputy Commander Ukrainian Army Major Serhij Heraimovich,

> Our soldiers participate in all MNB(E) training offered at Camp Bondsteel. Especially our medics are training on a regular basis with their American counterparts. There are also many professional development and language skills programs for our sergeants in the United States prior to being sent on this mission.[60]

This training and these bonds naturally help during joint actions:

Since arriving in the theater, the Ukrainian and American soldiers have conducted joint patrols. According to [Ukrainian] Senior Sergeant Andrej Chernata, these patrols have been very interesting: "Especially during joint patrols, we have the opportunity to share knowledge and our shared goals of security."[61]

Iraq.

The story of Ukraine-U.S. cooperation in Iraq begins with the deployment of the Ukrainian CBR-protection battalion to Kuwait during the active hostilities period of March-April 2003. The battalion was deployed by Ukrainian air transport to Kuwait 2 weeks after hostilities began.

This first deployment to the zone of hostilities near the Iraqi border proved several things. The problem of the individual equipment of Ukrainian personnel remained, as was the case in Kosovo, and the United States again provided support on the spot with some basic equipment including uniforms, protection gear, and footwear. But there was no doubt about the professionalism of Ukrainian personnel and the readiness of their combat equipment. In addition, the Ukrainians were located at the same base as the Americans, which added once again to their knowledge and appreciation of each other.

Though Ukrainian 19th CBR-protection battalion did not take part in active combat action, in August the third "special" battalion of Ukraine's 5th Separate Mechanized Brigade was formed from elements of this battalion. As far as follow-up deployment of the Ukrainian 5th Mechanized Brigade and higher staff elements to Iraq is concerned, the announcement was made in May 2003 during the conference in Warsaw on foreign troops' contribution to the stabilization operation. Though the news about the U.S. proposal to deploy a brigade headquarters and two battalions to Iraq (made to Ukraine earlier that month) was known previously, Ukraine's agreement still came as a surprise for many. When the Ukrainian proposal officially was made, according to witnesses, there was at first a moment of silence evidently caused by this surprise. Then some of those present started to ask their neighbors again to confirm whether they had understood the Ukrainian offer correctly. When it was confirmed by Ukrainian Deputy Chief of the General Staff Major General Oleg Sivushenko, the U.S. Marine officer present at the conference, who had just arrived from Iraq, could not hide his pleasure—"My people will go home!"[62]

After 3 months of intense preparations, in which U.S., Canadian, and British instructors took part, the Ukrainian contingent was fully deployed to Iraq in accordance with the schedule by August 17, 2003. Not everything went smoothly, and the heads of the Ukrainian MOD's armaments and logistic headquarters paid for this with their positions, as did several other high-ranking Ukrainian officers. Nonetheless, overall, after the previous deployment to Kuwait, this next deployment to Iraq once again proved the ability of the Ukrainians to deploy in time, relying primarily on their own airlift capability while partly using foreign sealift when necessary. The Americans, who pledged to compensate the Ukrainians for the transportation cost of both deployments—to Kuwait and Iraq—praised this ability.

The Ukrainian brigade was to replace the 3rd Marine Battalion at al-Kut, the capital of Wasit province in Iraq, southeast of Baghdad on a 140-km long sector of Iraqi border with Iran. The substitution was planned to be finished in 2 weeks. The first week, both the Ukraine and the United States had to accomplish missions together. During the second week, only instructors remained on the U.S. side to help Ukrainian personnel adjust. From early September, the Ukrainians assumed full responsibility for Wasit province's peace and order.

According to Ukrainian participant accounts,[63] as soon as the first elements of the 5th Brigade stepped onto Kuwaiti and then Iraqi soil, U.S. military personnel gave them comprehensive support. The United States helped in the organization of transportation, security, rations, deployment of personnel, and supply of the equipment for the Ukrainian military contingent. At briefings, as well as in everyday communication, the experienced U.S. troops continuously gave practical advice on how to operate in a hot climate, what to be careful of, what to do if someone gets sunstroke, how to maintain communications and orient oneself in an unfamiliar environment, how to act when under attack, etc.

At the Kut airfield, the command of the U.S. Marine battalion, which was to transfer control to the Ukrainian brigade, did everything to ensure that its comrades-in-arms would not be starting from scratch. Almost every day U.S. personnel met with personnel of the Ukrainian peacekeeping contingent and transferred their experience in performing their duties. Specifically, the 19th Separate Special Battalion, which was slotted to take over the defense of the Kut airfield, was given complete information about the most dangerous sectors of the perimeter of this large installation. The Marines told the Ukrainians about various incidents that had occurred over the 5 months since the operation began. They also shared their observations of the behavior of the residents of the city.

The specialists of the marine engineer unit were just as concerned with the health and lives of the Ukrainians. They collected and displayed a huge exhibit of the explosive devices that had been found on the former military airfield. The U.S. sappers fully characterized each mine, device, and grenade and recounted the story of an unfortunate incident involving two of their soldiers. At some points, Ukrainian and U.S. sappers had to work together to demine the territory of the airfield, which they did with understanding and trust in each other.

The company of the U.S. military police proved particularly helpful to the military police platoon of the Ukrainian brigade. The specific nature of assignments involving the patrol of the unknown city of Kut and the detention of law-breakers demanded special knowledge and practices. Thanks to the efforts of U.S. Captain Terry Dorn, Ukrainian peacekeepers were able to master the service quickly. Almost every night in the U.S. or Ukrainian headquarters, one could witness the examination and analysis of the joint operations in the city by this young woman. According to Ukrainian accounts, "Captain Dorn was concerned for the Ukrainians as if for her brothers. She wanted everything to turn out as it should and to avoid human losses."[64]

The joint cooperation of the Ukrainian and U.S. military, which began with transmission of operational experience, continued after the marines' departure, when the task of training an Iraqi territorial self-defense battalion was carried out jointly. The Ukrainians were responsible for selection and medical examination of Iraqi personnel, while training was organized together with U.S. instructors.

An interesting moment was recalled by a Ukrainian CIMIC officer. When head of the brigade's CIMIC section Lieutenant

Colonel Veleriy Kuzmin first approached local contractors, he was stunned by the previous practice. His U.S. predecessors allegedly were providing payments for orders without attempting to verify the local price for specific jobs and without any competitive bidding. This led to locals extracting large sums of money for rather simple and inexpensive orders. Therefore, in order to reduce waste, he had to exert strong efforts and experience conflicts to break the habits of many local contractors spoiled by U.S. generosity.[65] This is an interesting "role reversal" for the United States and Ukraine, considering the complaints by Americans that the Ukrainians made the same errors and encouraged "price-gouging" in the early days of U.S.-Ukraine military contacts (see Part II). It appears that the Ukrainians had paid attention to the "lessons learned" from that earlier military cooperation, and as a result may actually have had more experience and knowledge on this subject than the U.S. units involved in this particular operation.

Another example, which might have something in common with the internal U.S. debate about the "tough" Army versus "liberal" Marine approaches to local population in the occupied territories,[66] was observed in the Iraqi town of Suwayrah by a *Los Angeles Times* correspondent:

> . . . Some residents are happy that the Ukrainians have taken over from the Americans, who they complained insulted residents and showed disrespect to Iraqi women when raiding homes or conducting body searches. "The Ukrainians treat us in a very nice way, completely different from the Americans," said Adnan Hamid Abbas, a lawyer. "They never shout at us." But others said the Ukrainians' easygoing nature meant they were not as aggressive as the Americans in eradicating supporters of Saddam Hussein or resistance fighters who were staging attacks in the area. "The Ukrainians are cowards, while the Americans are tough," said Ghasan Ali Izzi, a television shopkeeper. "I prefer the Americans."[67]

Whatever the pros and cons, perhaps an analysis of the more "liberal" Ukrainian approach may yield lessons applicable to the internal U.S. debate.

Outside the Ukrainian 5th Brigade, the officers of the Ukrainian staff element at the higher "Center-South" division headquarters and at the Coalition headquarters, who were working together with

Americans, Poles, Spaniards, and other allies, on many occasions recognized the valuable experience they received during joint peacetime training, joint peacekeeping missions, and at the courses for multinational staff officers in Kiev. This experience allowed them to be interoperable with their U.S. partners and other nationals in Iraq.

Regarding the preliminary analysis of the first rotation of the Ukrainian brigade to Iraq, the United States appeared basically satisfied, based on the overall ability of Ukrainians to deploy to the theater of operation and perform the assigned mission. However, honest feedback from Polish partners—who have had the opportunity to cooperate closely with the Ukrainian brigade on an everyday basis within a joint formation—has been more specific and more critical, particularly concerning the low level of English proficiency among Ukrainian officers in Iraq and the slow pace of forming and training the brigade to the required standard.[68]

During the second rotation of Ukrainian troops in Iraq, when 6th Brigade was substituted for the 5th Brigade in February-March 2004, the events in Wasit province, as everywhere in Iraq, took more dramatic turn. In addition to routine instances of cooperation of the U.S. and Ukrainian sappers, medics, logisticians, etc., several combat engagements took place. At the start of clashes with the militia of Shiite cleric Muqtada-al-Sadr, on April 6-7, two Ukrainian mechanized platoons for 2 days defended the Coalition Provisional Authority (CPA) building at the capital of Wasit province, the city of Kut, and another Ukrainian mechanized platoon defended the bridge over the Tigris River. They withstood a 4-hour long attack by overwhelming numbers of Sadr militia and further attempts to blow the CPA administrative building, as well as continuous mortar and RPG attacks. They suffered one dead and five wounded, while killing over a hundred attackers. Two U.S. F-16s and later two *Apache* helicopters came and left without firing a shot or launching a missile apparently because of problems with target identification and lack of coordination with Ukrainians.

The Ukrainian detachment managed to evacuate safely 37 U.S. and 6 Polish civilian personnel under their protection from the siege. According to Ukrainian official sources, this was done at the request of U.S. civilian CPA personnel. Two U.S. *Apache* helicopters covered the retreat.

Unfortunately, many U.S. newspapers reported the incident in slightingly negative tones, as if Ukrainian troops simply abandoned the city of Kut (some 250,000 inhabitants!) to the Sadr militia. In fact, it never was a mission of the lightly armed Ukrainian detachment to hold the city. Their mission was to do exactly what they had done—to provide for security of civilian administration—which they accomplished with tactical brilliance. Law and order at Kut were restored soon after by the U.S. combat brigade, so in the aftermath, it is perfectly clear that, opposite hasty media accusations, there was no need for the Ukrainian company to replicate another Stalingrad and risk the lives of civilians and their own.

U.S. Secretary of State Colin Powell thanked Ukrainian allies in his letter dated April 9 to Ukrainian Foreign Minister Kostiantyn Hryshchenko. Powell highly praised the courage and bravery shown by the Ukrainian troops, particularly during the events in the city of Kut. During retaking control of Kut, the Ukrainian battalion stationed near the city performed a supporting role and later renewed patrolling the city and conducting searches. For instance, during the first day, they detained three militiamen pretending to be fire-fighters and discovered a large weapons cache. The other two Ukrainian battalions from the 6th Brigade were not engaged in serious combat: one was doing border guard duty defending the 80 miles long border with Iran (they also had seizures of large amounts of weapons), and the third battalion was busy with de-mining, patrolling, and CIVIC missions. Associated Press reported the events:

> U.S. troops drove into Kut before dawn Friday, pushing out members of the militia headed by radical Shiite cleric Muqtada al-Sadr that had seized the southern textile and farming center this week after Ukrainian troops abandoned the city under heavy attack. . . . Brigadier General Mark Kimmitt said he expected the operation to retake Kut from al-Sadr's al-Mahdi Army militia would be finished by Saturday morning. . . . Kimmitt told CNN he believed there were 300-400 al-Sadr militants in Kut on Thursday night who had been trying "to intimidate the people" in the city of about 250,000. The Kut operation represented a major foray by the American military in a region where U.S. allies have struggled to deal with the uprising.[69]

However, the attacks on Ukrainian troops in Wasit province continued, and on April 19, a Ukrainian patrol on three armored personnel carriers was attacked by a large group of terrorists, which detonated two roadside bombes and opened fire from RPG and small arms. The Ukrainians returned fire, killing five and wounding seven. In another similar ambush on April 28, 2004, the Ukrainian patrol was less fortunate, when two Ukrainians died and five were wounded.

But in the latter case, as in other similar cases, U.S. MEDEVAC helicopters and U.S. medical personnel did everything possible to save wounded Ukrainians. U.S. attack helicopters always were ready to provide fire support. However, certain problems of coordination between Ukrainian and U.S. troops became evident, so in the aftermath of the April-May events in Wasit province, U.S. CENTCOM supplied the Ukrainian brigade with an additional number of U.S. radios, which Ukrainians had in limited quantity.

At home, Ukrainian leftists immediately accelerated campaigning on the withdrawal of Ukrainian troops from Iraq. This campaign, along with absence of a balanced and positive coverage in foreign press, made it more difficult for Ukrainian authorities to explain to the public why Ukrainians are dying in Iraq.

Again, as in the case of events in Kosovo in February 2001, the lack of appreciation by the U.S. media was perceived by Ukrainians as reluctance to note their important role in dramatic April-May 2004 events in Iraq. Overall, Ukrainian participation in the stabilization operation in Iraq, as previously in Kosovo, has provided a new opportunity to gain real-life experience, to learn from each other, and to test the value of peacetime engagement. These ultimate tests of doctrine, training, and equipment generally have shown that, while Ukraine is capable of consistently producing ad hoc successes, a more systemic approach is needed to develop a stable, long-term capability for deploying interoperable units. The repetition of previously recognized problems strongly suggests that interoperability issues should be included in a system of regular review of operations, training, and doctrine—that is, a mandatory part within a more specific and effective Ukrainian "lessons learned" system.

PART V

CONCLUSIONS

The principal U.S. approach to cooperation with Ukraine, a post-Soviet nation with long and complex history, has consistently been to help in building a stable, prosperous democracy that can become a viable economic and security partner to the West. An important role in these efforts belongs to military cooperation: within the bilateral military-to-military contacts programs, within NATO partnership events, and through the practical accomplishment of peacekeeping and humanitarian missions.

In the course of 13 years of cooperation, Ukraine and the United States have gone through periods of cautious rapprochement, inflated expectations of "strategic partnership," and sober reevaluation. Ukraine's recent ambitious declaration of intent to become a NATO member was welcomed by the United States, although cautiously, given that Ukraine has not been very successful in building a firm democratic foundation and conducting defense reform.

Although the search for the most appropriate political *modus vivendi* for bilateral relations still continues, the military dimension of the relations between two countries has always remained cooperative. Most important is that in the military sphere there are no insurmountable ideological, geopolitical, or cultural differences between Ukraine and the United States. The history of military cooperation has proven that, despite Ukraine's many political and economic problems, as well as those of a cultural and military nature (bureaucracy and over-centralization, Soviet legacy of equipment and doctrine), certain core interests provide firm ground on which to continue mutually advantageous military cooperation. These core interests are, at their most basic, U.S. willingness to support the preservation of Ukraine's independence as a key to regional security and Ukrainian willingness to cooperate with the United States in fighting terrorism and preserving international peace.

The two countries have developed elaborate cooperative mechanisms, which permit rather effective implementation of joint events. The Ukrainian military appears genuinely to be interested in this cooperation, is generally technically and intellectually capable, but is still a rather long way from compatibility with U.S. cultural and doctrinal standards. But if we consider the starting point, the results are impressive.

At the start of their cooperation, there was practically no ground to talk about interoperability in the *traditional* sense between U.S. and Ukrainian (post-Soviet) militaries.[70] But 13 years of military cooperation have allowed for achieving certain limited progress in major interoperability areas between the U.S. and Ukrainian militaries. As a result, the relationship has grown from simple peacetime engagement to conducting successful combined peace-keeping operations.

As has happened so far, it is exactly in the area of peacekeeping where the United States has needed—and will continue to need—the Ukrainian military the most. Bosnia, Kosovo, Afghanistan, and Iraq have shown that the United States is capable of winning regional wars without decisive support from its allies. But these campaigns also have proved that the United States has significant limits in providing for post-conflict resolution (peacekeeping and peace-building) without support from allies, even those as distant as Ukraine. This is very important, since no war can be considered victorious if the post-war situation deviates too far from prewar objectives.

For the United States, the experience of military cooperation with Ukraine has proved that U.S. military contact programs and peacetime engagement are a good way to understand the people with whom you are engaging, and evaluating whether they are "really on your side." Quite likely, this experience also proved for the United States an already known classical virtue of coalitions: they allow smaller nations to feel important, while they allow stronger nations to consider that others share the burden.

For Ukraine, military cooperation with the United States has provided many opportunities for expediting reforms in the security sector. Unfortunately, Ukraine has not been very successful in using these opportunities. In particular, this is the result of the lack of strong political direction and sufficient funding, compounded by the failure of Ukraine's military to introduce a system to process the lessons learned effectively. To alleviate partly the impact of Ukrainian problems, the country could have done better in developing its own "U.S. specialists" to take some of the burden off the United States; that is, to train enough "plug and play" liaison officers.

If, for the sake of comparison, a third party—for example, new NATO member and Ukraine's neighbor, Poland—is used as an indicator, the general conclusion would be that Ukrainians, as a fighting force, are not more deficient than Poles. Rather, Ukrainians are less interoperable in terms of language, doctrine, and equipment. However, Ukraine has some unique capabilities it can provide, such as airlift, missile/space, radars, tanks, CBR-testing and protection equipment, and other high-tech possibilities.

Ukraine, indeed, has a lot of assets potentially to contribute to combined operations with the United States, but the challenge still remains how better to make them interoperable. In answer, the results of this study generally point to the need for a two-tier approach to interoperability: the first tier being continued efforts to develop **compatible capabilities for the low intensity conflict** (peacekeeping); the second tier being the identification and improvement of **complementary**—rather than comparable—**capabilities for high intensity conflict**.

At this moment, however, because of the number of political and security reasons indicated above, in practical terms it is relevant to speak primarily about the value of interoperability in the low intensity conflict. To ensure continued success, **more systemic approach to U.S.-Ukraine military relations should be recommended.**

First, given the total domination of the Army agenda in bilateral military programs, consider shifting from the current practice of appointing the U.S. Defense Attaché in Kiev from the U.S. Air Force to more relevant and logical representation—from the U.S. Army. For the same reason, Army program events should become undisputed priorities of the Program of Bilateral U.S.-Ukraine Military Cooperation.

Second, in order to develop a stable, long-term capability for deploying interoperable Ukrainian units, a lot has already been done and is currently planned to be done. But additional efforts are still needed. These are:

- Developing effective Ukrainian "lessons learned" systems, in which interoperability issues should be included in regular review of operations, training, and doctrine.
- Initiating the comprehensive and systemic adaptation of Ukrainian manuals, headquarters' techniques, and leadership styles to U.S. and NATO models.
- Supporting Ukrainians in adapting wargaming techniques to provide for interoperability at the strategic-operational levels.
- Providing more focused support to the Ukrainian side in training operational officers capable of pursuing interoperability issues. To this end, establish the permanent placement of the U.S. instructors at the special courses for the officers of multinational staffs within the National Defense Academy of Ukraine.
- Selecting and training Ukrainian instructors for Ukraine's National Academy of Defense and other relevant training facilities in view of their contribution to building interoperability.
- Helping transition from the current Ukrainian practice of creating ad hoc units for missions abroad to deploying regular units, first of all, from Rapid Reaction Forces.
- Providing targeted support in equipping Ukrainian Rapid Reaction Forces with interoperable command, control, and communication equipment.
- Focusing joint training exercises on actual units, which will deploy out of country.

Third, recognizing that success of Ukraine's efforts in reforming its military, particularly in increased interoperability, depends to a significant extent on success of wider governance reform. Ukraine should consider more targeted efforts in training Ukrainian defense experts from the staffs of the Parliament, the Cabinet of Ministers, Administration of the President, National Security and Defense Council, Ministry of Foreign Affairs, Ministry of Finance, etc.

As proved by the experience of practical missions together in peacekeeping, human and intellectual factors (an understanding of each others' national and military cultures, common language) are more important than technical systems. Thus for low intensity conflict, "operational interoperability" (similar doctrines, planning methods, training, and basic doctrinal/cultural understandings) appears to be more important than "technical interoperability."

But by deploying a brigade to Iraq, Ukraine has raised the level of its cooperation to a new height, which will, in turn, be a new test to the value of peacetime military cooperation. If, despite all conceivable political and military problems, this new level of cooperation is successful, it might open the door for partnership beyond peacekeeping.

Thus, there are grounds to think that options for greater interoperability for higher-intensity operations should be considered as well. This analysis proves that the opportunities are there—but for these to materialize, both countries' militaries will need to continue efforts to further strengthen the common capabilities and bonds that U.S.-Ukraine military cooperation has already helped to build.

Overall, U.S.-Ukrainian military cooperation has developed more slowly and less efficiently than most U.S. participants have expected. But given the magnitude of the nation-building challenge for Ukraine, it could be assessed as a qualified success. This cooperation has brought tangible results for both sides, as well as valuable lessons for modern relations between the United States and post-totalitarian states.

ENDNOTES

1. For more details, see Vladimir Bezkorovainy, "Navy and Maritime Policy of Ukraine," *Defense Express—Arms Export and Defense Industrial Complex of Ukraine: News and Analysis*, No. 11, 2003, p. 8.

2. "It was intellectually decapitated as a matter of deliberate policy during the Stalin years and beyond so that the most able and energetic Ukrainians were either killed or magnetically attracted to Moscow and Russified." Zbigniew Brzezinski, "Ukraine and Europe," *Fulfilling the Promise: Building an Enduring Security Partnership Between Ukraine and NATO*, Ashton B. Carter, Steven E. Miller, and Elizabeth Sherwood-Randall, eds., The Stanford-Harward Preventive Defense Project, 1999, pp. 33-34.

3. For a comprehensive account of the Soviet military legacy in Ukraine, see Oleksiy Melnyk, "Ukraine's Cold War Legacy 12 Years On: A Burden from the Past, a Problem for the Future," in Alyson J. K. Bailes, Oleksiy Melnyk and Ian Anthony, *Relics of Cold War (Europe's Challenge, Ukraine's Experience)*, SIPRI Policy Paper No. 6, November 2003.

4. Besides those in the armed forces, Ukraine had a large number of troops in other military formations: 130,000 troops under the Ministry of Internal Affairs, Border Troops, etc. Some 12,000 officers and NCOs of non-Ukrainian origin left Ukraine, while 33,000 came back to Ukraine from other former Soviet republics. For more details, see Andreas Heinemann-Gruder, "Becoming an Ex-military man: Demobilization and Reintegration of Military Professionals in Eastern Europe," *BICC Brief 26*, August 2002, p. 18.

5. Future Ukrainian Joint Rapid Reaction Forces are expected to include land component (three airmobile brigades and various supporting units), air component (three squadrons of MIG-29 *Fulcrum*, SU-25 *Frogfoot*, and IL-76 *Candid* each), and Naval component (three ships: command ship, frigate, and a large landing ship). Its total strength is expected to number around 30,000.

6. The course employs instructors from NATO nations and is open to NATO partners.

7. "Remarks by the President in Address to the Supreme Soviet of the Ukrainian Soviet Socialist Republic," The White House: Office of the Press Secretary, August 1, 1991. Cited from Sherman Garnett, *Keystone in the Arch: Ukraine in the Emerging Security Environment of Central and Eastern Europe*, Washington DC: Carnegie Endowment for International Peace, 1997, p. 4.

8. Roman Popadiuk, *American-Ukrainian Nuclear Relations*, McNair Paper 55, October 1996, Washington, DC: Institute for National Strategic Studies, National Defense University, p. 60.

> One also cannot dismiss the cultural influences that Washington had to deal with. Having been a submerged nation for centuries, there was limited appreciation in Washington of the historical and cultural

> motivations in Kiev. There was a learning process for Washington policymakers, whereby outside the words of recognizing Ukraine as an independent state, they had to come around to dealing with and treating it as an independent state, and this meant respecting its positions and negotiating rather than dictating or expecting automatic action.

9. Sherman Garnett, "Ukraine's Regional Security Challenges," *National Security & Defence*, No. 7, 2000, p. 19.

> . . . during the negotiations with Ukraine on nuclear issues, one of the biggest problems on my side was assuming that the Ukrainian bureaucracy was like us. . . . I would go to U.S. Government meetings on Ukrainian nuclear weapons, and there could be 75 people. These are policy questions, big decisions. The U.S. President made those decisions himself, but there were 75 people who wrote papers and participated in this process. There were not that many people on the Ukrainian side. There is a difference in bureaucratic structure, culture and outlook.

10. "We will sustain our effort to help integrate Ukraine more fully into international institutions and structures. . . . an independent, unitary, secure, democratic, prosperous, self-confident, integrated Ukraine is a keystone in the architecture of this new Europe." Strobe Talbott, "The New Ukraine in the New Europe," *Fulfilling the Promise: Building an Enduring Security Partnership Between Ukraine and NATO*, p. 22.

11. Sherman Garnett, "Ukraine: Strategic Partner or Strategic Problem?"—*Towards a New Ukraine III, Geopolitical Imperatives of Ukraine: Regional Context*, Theofil Kis, Irena Makaryk, Natalie Mychajlyszyn, eds., with Irena Bell, Ottawa, Canada: Chair of Ukrainian Studies, University of Ottawa, 2001, p. 165.

12. On February 3, 2003, President Kuchma in his address to the diplomatic corps in Kiev underlines, "In its relations with the United States, Ukraine remains loyal to the ideals of strategic partnership in spite of the existing problems. I want to stress with all confidence that we have never intended and do not intend to revise our policy toward the United States." See Embassy of Ukraine in the United States, *www.ukremb.com.*

13. Ambassador Herbst, Speech on Democracy in Ukraine, Institute of International Relations, Taras Shevchenko National University, Kiev, December 23, 2003, *http://kiev.usembassy.gov/files/031223_herbst-iir-remarks_eng.html.*

> As Americans, we would like to see democracy flower here for two reasons. First of all, the U.S. and Ukraine share many common interests. We would like to establish a deep, long-term friendship and more, but that can only happen if Ukraine is a true democracy. Secondly, we would like to see long-term stability in Ukraine and history teaches that democracy, well grounded in the ethos of an open society, is the best way to ensure both enduring stability and felicity. That in turn will contribute to European security and stability.

14. Peter Eisler, "U.S. helps Ex-Soviet states scrap weapons," *United States TODAY*, January 31, 2003, *http://www.United Statestoday.com/news/world/2003-01-30-ukraine-United Statest_x.htm*.

15. U.S. Government Assistance to and Cooperative Activities with Eurasia, FY 2002, Ukraine. Released by the Bureau of European and Eurasian Affairs, January 2003, *http://www.state.gov/p/eur/rls/rpt/23629.htm*.

16. "CINC" is an abbreviation that once denoted a regional U.S. "Commander in Chief"; a U.S. regional military commander is no longer referred to by this term, but simply as a "Combatant Commander," or "Commander, USEUCOM" for example.

17. *Ibid*.

18. According to the Office of Defense Cooperation of the U.S. Embassy in Ukraine.

19. James Sherr, "Iraq and the impermissibility of failure," *Dzerkalo Tyzhnia*, October 25-31, 2003.

20. The Joint Contact Team Program (JCTP) was established by General Colin Powell in 1992 in an effort to promote regional stability in the newly emerging democracies of Central/Eastern Europe by presenting them with the U.S. example of how a civilian-controlled military works in a democratic free-market orientated society. Today, the JCTP is a pillar of USEUCOM's security cooperation activities in Central/Eastern Europe and the Trans-Caucasus region. Through orienting host nation militaries to DoD-wide organizations, programs, activities, and standards, the program supports USEUCOM theater objectives of stability, democratization, military professionalism, closer relationships with NATO, and, for the new NATO members, preparing for NATO integration.

21. The National Guard State Partnership Program was established in 1993 in response to the radically changed political-military situation following the collapse of Communism and the disintegration of the Soviet Union. Today 39 U.S. states, 2 territories, and the District of Columbia are partnered with 44 countries around the world. SPP is also an added dimension of JCTP. The SPP encourages the development of long-term institutional and personal relationships and allows more Americans to become directly involved in assisting these countries as they transition to a democratic society. Reserve components conduct approximately 25 percent of JCTP events.

22. Stephen D. Olynyk, "Ukraine as a Post-Cold War Military Power," *Joint Force Quarterly*, Spring 1997, p. 88.

23. For a good analysis of Ukrainian problems with strategic guidance documents and other key issues of current Ukrainian security and defense reform, see James Greene and Christopher Donnelly, "The Basic Principles of Structuring Modern Armies—in the Context of Ukrainian Security & Defense Reform," *National Security & Defence*, No. 7, 2003, pp. 54-61.

24. See the *National Military Strategy of the United States of America*, 1995.

25. See the *National Security Strategy of the United States of America*, September 2002.

26. Colonel (U.S. Army retired) Harry Simmeth authored the most complete account of the 1995-99 period of the U.S.-Ukrainian military cooperation in his article, "U.S.-Ukraine Military Cooperation and Strategic Partnership," *National Security & Defence*, No. 12, 2000, pp. 74-80.

27. *Ibid.*, p. 76.

28 ". . . The Military Liaison Team is a comparatively inexpensive project, costing the U.S Defense Department around $400,000 annually for office upkeep, travel, and staff salaries. It is a price the U.S. and NATO are willing to pay in return for the Ukrainian military's help in maintaining peace." See Katya Cengel, Kyiv Post Staff Writer, "State wants to take a bite out of military might," *http://www.calguard.ca.gov/ia/Military/UMOD%20-%20further%20downsizing.htm.*

29. Harry Simmeth, "U.S.-Ukraine Military Cooperation and 'Strategic Partnership'," p. 78.

30. Timothy Shea, "Shaping on NATO's Doorstep: U.S.-Ukraine Relations," *Joint Force Quarterly*, Autumn/Winter 2001-02, p. 64.

31. *Ibid.*, pp. 62-63. For more of critical insight to the problems of U.S.-Ukrainian military cooperation, see Timothy Shea, "Post-Soviet Maskirovka, Cold War Nostalgia, and Peacetime Engagement," *Military Review*, May-June 2002.

32. See Jim Garamone, "U.S., Ukraine Sign Annual Plan of Cooperation," American Forces Press Service, Brussels, Belgium, December 6, 2000.

33. Author's interview with Major Ostrom on November 19, 2003.

34. In addition to these major areas, many other important factors influence the level of interoperability such as force structures, recruitment, security arrangements, and cultural traditions (casualties sensitivity, food, religion, and social status).

35. Ukrainian MOD statistics.

36. In early 2001 NATO and Ukraine took stock with the "Language Training Pilot Project." According to the project report recommendations, Ukraine did "need to direct military cooperation programs not just toward providing language training, but also toward improving the capabilities of Ukraine's own structures to plan, conduct, and manage language training programs."

37. Victor Hudym, "Ukraine's experience of participation in KFOR peace-keeping operation and in PfP program," *Peacekeeping activity of Ukraine: cooperation with NATO and other European security structures*, Joint Publication of the National Institute for Strategic Studies (Ukraine) and the Royal Institute of International Affairs (the UK), Kiev, 2002, p. 82.

38. "In the spirit of PfP" is a term typically applied to an exercise or event not sponsored by NATO but in which the participants generally follow NATO

guidance and procedures. It is intended to go somewhat beyond familiarizing the participants with each other's methodologies to the level of working together in a NATO-like environment.

39. Data from the press release of the briefing of the Army of Ukraine Commander-in-Chief Colonel General Petro Shuliak, May 26, 2003, *http://www.mil.gov.ua/index.php?part=news&sub=read&id=2239.*

40. Major General Nicholas Krawciw, U.S. Army, Ret., "Peacekeeping and Joint Military Operations: Lessons Learned and Next Steps," *Fulfilling the Promise: Building an Enduring Security Partnership Between Ukraine and NATO*, p. 50.

41. *Defense Analysis Report on the Armed Forces of Ukraine*, prepared by The Office of the Assistant Secretary of Defense for International Security Policy and The United States European Command, October 24, 2002, p. 23.

42. See "Complete "Internationalisation," *Narodna Armiya*, December 2, 2003.

43. For details, see "A group of cadets from the U.S. Military Acedemy West Point Study in Odessa Army Institute," Press release of Ukraine's MOD, March 18, 2003, *http://www.mil.gov.ua/ukr/index.cgi?read_news.htm&Uk&1992*; "As Part of the Ukrainian-U.S. Program of Military Contacts, American Officer Served in Airmobile Unit of the Armed Forces of Ukraine," Press release of Ukraine's MOD, January 21, 2003, *http://www.mil.gov.ua/ukr/index.cgi?read_news.htm&Uk&1784.*

44. Harry Simmeth, "U.S.-Ukraine Military Cooperation and 'Strategic Partnership'," p. 80.

45. Krawciw, p. 53.

46. Quote supplied by COL Simmeth from record of a meeting with a Ukrainian delegation at the Pentagon in 1996.

47. The author of the monograph, as a graduate of the U.S. Army War College and participant of many bilateral cooperative events, has utilized and popularized his knowledge of U.S. doctrine in a number of publications on the issues of personnel management, leadership, reserve component, urban warfare, civil-military relations, peacekeeping, and transition to an all-volunteer force. Many other Ukrainian IMET Program graduates utilized their knowledge of the U.S. doctrine in their respective publications and in practice of their assigned duties.

48. In practical terms, the most significant influence on Ukraine's doctrinal development, beyond the United States, was exerted by Canada, Germany, and the UK.

49. "Military Cooperation between Ukraine and the USA remain one of the key directions of Ukraine's Armed Forces international activity," Press release of Ukraine's MOD, December 3, 2003, *http://www.mil.gov.ua.*

50. "Issues of organizing Ukraine's Navy command and control and logistical support in accordance with NATO standards will be discussed during the visit of the delegation of NATO Supreme Allied Command Atlantic," Press release of Ukraine's MOD, July 21, 2003, *http://www.mil.gov.ua/ukr/index.cgi?read_news.htm&Uk&2425.*

51. James Greene and Christopher Donnelly, "The Basic Principles of Structuring Modern Armies—in the Context of Ukrainian Security & Defense Reform," *National Security & Defence*, No. 7, 2003, pp. 54-61.

52. U.S. Government Assistance to and Cooperative Activities with Eurasia, FY 2001, Ukraine, released by the Bureau of European and Eurasian Affairs, March 2002, *http://www.state.gov/documents/organization/17714.pdf*.

53. U.S. Government Assistance to and Cooperative Activities with Eurasia, FY 2002, Ukraine, released by the Bureau of European and Eurasian Affairs, January 2003, *http://www.state.gov/p/eur/rls/rpt/23629.htm*.

54. "In the near future, Ukraine intends to resolve the problem of secure communication of its NBC-protection battalion in Iraq," Defense-Express, December 23, 2003, *www.defense-ua.com*.

55 . Krawciw, p. 50.

56. Monika Blikes, "KFOR Secures Kosovo Elections," *KFOR Chronicle*, October 28, 2000, *http://www.nato.int/kfor/chronicle/2000/nr_001028.htm*.

57. *Ibid.*

58. Dmitri Shkurko, "The war in Kosovo has flared up. This Time It Affected Ukrainian Peacekeepers in Serbian Enclave," *Narodna Armiya*, February 23, 2001.

59. Kevin Dougherty, "In Wake of Bus Bombing, American's Efforts Have Earned Them Serb's Trust," *Stars and Stripes*, February 21, 2001.

60. Veronica Saffo, "Ukrainian Patrol," *KFOR Chronicle*, April 18, 2003, *http://www.nato.int/kfor/chronicle/2003/chronicle_04/18.htm*.

61. *Ibid.*

62. From interview with Warsaw conference participant, who preferred not to be identified.

63. Andriy Lysenko, "Common Goal United All Militaries," *Armiya Ukrainy*, December 4, 2003.

64. *Ibid.*

65. Andriy Lysenko, "'CIMIC'—a New Page of Post-War Diplomacy in Iraq," *Narodna Armiya*, January 6, 2004.

66. For details, see Thomas E. Ricks, "Marines to Offer New Tactics in Iraq; Reduced Use of Force Planned After Takeover From Army," *The Washington Post*, January 7, 2004.

67. Richard C. Paddock, "Wounding of Ukrainians Could Give Other Nations Pause," *The Los Angeles Times*, October 31, 2003.

68. Lessons-learned Conference, Polish Embassy, Kiev, December 2003.

69. "U.S. Forces Retake Most of Key Iraqi City," *The Associated Press*, April 9, 2004.

70. Ralph Peters, Letter to the author, October 24, 2003.

The [U.S. and Soviet prior to 1991] systems were so profoundly different that the only interoperability basis is simply that all are soldiers, with some shared general knowledge and purpose, subject to discipline, etc. But the functional systems and approaches to leadership and personnel management are profoundly different. . . . I just do not see any doctrinal common ground, whatsoever. . . as regards the present situation. I expect a great many lessons have been learned, on both sides, from interoperability requirements in the former Yugoslavia and elsewhere.

APPENDIX

KEY INTERVIEWS

Major General (U.S. Army Ret.) Nicolas Krawciw,
Special Assistant to the Secretary of Defense for Military-To-Military Relations between USA and Ukraine

The principal American interest in Ukraine, for obvious regional security and economic reasons, is in enhancing its independence and democratic institutions. I believe that, for similar reasons, Ukraine's best interests are to become part of Europe.

Our current status of military cooperation with Ukraine became somewhat stalled following the "Kolchuga" disclosures, but is reviving as Ukraine, with its brigade in Iraq, is providing valuable professional assistance. We are working with NATO HQ and with other NATO countries to assist in Ukraine's preparation for the MAP process. Senior American government officials are concerned that some nasty aspects of Ukrainian election year power politics may negatively influence NATO allies in their deliberations in preparation for the Istanbul NATO Summit concerning the issuance of an invitation to Ukraine to join NATO.

Not much has been achieved in technical military cooperation. After long delays, the classified information agreement between the United States and Ukraine has been signed last year. In time, if relations remain on course, this may bear fruit in some meaningful technical and technological cooperation.

Concerning cultural aspects of interoperability between our militaries, I believe much has been achieved. We understand and appreciate each other better. The intrinsic value of our exchanges I describe in the second part of this message. I would like to add that even the high level visits and exchanges have contributed to our mutual understanding of our respective capabilities, methods, politics, etc. Changes in senior Ukrainian military leadership mindsets concerning threat definition, management, leadership, military reform, and care of people have been progressive, visible, and contribute to "cultural interoperability." The best examples that I can think of are these:

a. When we first started the training exercises in Yavoriv and in the Black Sea, Ukrainian units were focused not so much on training to achieve the training objectives of the exercise but to look good in various demonstration vignettes at the "Opening Day," or "VIP Day," or "Closing Day" when various high level visitors were expected to attend. By 2000, this began to be changed to a focus on exercise training objectives and visitors were shown what is actually being achieved.

b. Similarly, discussions of defense reform over the last few years turned from showcase briefings to discussions of what is feasible, possible, and achievable based on available budgetary funding.

Events of this year [2004] will determine the future course of our cooperation. In any case, we on the U.S. side will continue our assistance to meaningful Ukrainian defense reform for as many years as is necessary and is desired by Ukraine.

I would also provide comments on the main differences between the U.S. and Ukrainian militaries. During the late 1970s, the U.S. Army began a serious analysis of Soviet how-to-fight doctrine. We learned that Soviet highest level commanders had a significant amount of freedom of action at the strategic-operational level, but less at the operational-tactical level, and very little at the tactical level. We also knew of the impressive Soviet artillery and combat engineer capabilities. Our post-Vietnam reform became oriented on improvement of these two capabilities to try to match those of the Soviet Union. However, at the operational tactical level, we felt that we had the advantage in the way that our lower level commanders (division down to battalion) planned and executed operations. The system that evolved in Vietnam gave these commanders areas of operation and general missions but would let them develop and coordinate plans and actions within the assigned areas. That was extended down to company level. Needless to say, that approach stimulated individual initiative and contributed to leadership development in the post-Vietnam period.

When we began our Partnership for Peace and other bilateral exchanges with the Ukrainian Armed Forces in 1994, it became clear that many of the Soviet practices and organizational concepts that still existed, while possibly useful for a major continental land war, were not suited to small regional conflicts or to peacekeeping. There were no planning staffs at regimental or battalion levels. Everything was being directed from the top. Colonels, lieutenant colonels, and majors had very little leeway to change training schedules or to influence organizational requirements. Equipment looked good, but there were no funds to exercise or maintain it. We were told that, due to a general lack of funding, force modernization apparently stopped in the early 1980s. There was no real NCO corps; officers were performing tasks of seargeants. Inspections by higher commanders or by the Inspector General were dreaded and not really systemic. On the other hand, Ukrainian officers were highly educated; many showed a genuine interest in different ways of doing things.

Beginning in the summer of 1995, our "Peace Shield" exercises in Yavoriv were designed to share our tactical and operational methodologies with the Ukrainian Ground Forces (Ukrainian Army) and with other participants. Similarly, the "Sea Breeze" Exercises in 1997 and 1998 were intended to assist the Ukrainian Navy to develop its sea and shore based staffs. Most importantly, Ukrainian officers trained in the United States (IMET) and in other NATO countries were bringing back to Ukraine better examples of various military methodologies.

In my estimation, over the years, IMET, similar education in other countries, actual peacekeeping operations, and various combined field or staff exercises, provided the Ukrainian Armed Forces with various models of operational/tactical techniques more suited to current real world situations and exposed Ukrainian participants to the more open ways of conducting military affairs that exist in the West. By the summer of 2000, we could see that certain directorates of the Ukrainian Ministry of Defense began to develop methodologies for defense reform. In some cases, particularly in military education, reforms started taking place. Plans were drawn up to develop an all-volunteer enlisted force, with an NCO Corps at its core. As of the end of last year, while budget and legislation dependent, the reforms that are taking place; and those that are contemplated, particularly the drawdown to much lower force levels and the restructuring to brigade organizations, seem to be on the right track if Ukraine really desires eventual NATO coalition security. We are now poised to assist the Ukrainian Armed Forces in their preparation for participation in NATO's MAP once Ukraine receives NATO's invitation for membership.

Higher morale, esprit de corps, lower level leadership, and initiative will develop as the economy continues to improve, as military personnel get better pay, and as political turbulence diminishes. The ethical grounding of most Ukrainian officers is solid and will surface when the right political and senior military leadership is in place. Examples of individual integrity at the highest levels will assist in the grooming of better leadership at all levels of command.

Colonel (U.S. Army Ret.) Harry Simmeth,
former Branch and Division Chief, Joint Staff J-5 Strategy, Plans and Policy; and former Commander, USA Opposing Force (OPFOR) Regiment, U.S. National Training Center, Fort Irwin, CA.

It is difficult to pass a summary judgment on the overall value of this cooperation, as it is subject to analysis on a number of levels. For example, it clearly can be said to have paid large dividends in the sense of preparing U.S. and Ukrainian forces to work together in real-world contingency operations, e.g., Bosnia, Kosovo, and now Iraq. However, part of the U.S. (and NATO) rationale in conducting this cooperation was to encourage greater reform in the Ukrainian armed forces. The latter effect, it seems to me, has been minimal overall. If you want me to categorize it as "good or bad" overall, however, I would have to say "good," but I am disturbed at aspects of the "negative side" of the equation.

For example, there still exists a tendency in the Ukrainian Defense bureaucracy to overcentralize planning and decisionmaking. In any system, this inevitably leads to various inefficiencies and frustrations.

There was a period of time—I would estimate about 1997-99—during which I sensed a serious effort on the part of certain Ukrainian military leaders to correct this situation. One of the more positive outcomes of this tendency was the creation of the U.S.-Ukraine "Colonels' Conference," a mechanism through which joint and Service Colonel-level and below planners from each country met regularly to set objectives and priorities and formulate proposed schedules of events and programs. A true test of this mechanism arose in 1998, when the Colonels' Conference recommended significant modification of a new program proposed jointly by the Ukrainian Defense Minister and a U.S. Deputy Undersecretary of Defense. The Colonels collectively felt that the funds involved could be better utilized on a similar existing program rather than creating a new, duplicative one. To the U.S. Colonels, this was considered "advice" to be either heeded or overruled; the Ukrainian Colonels, although in complete accord, seemed to feel the Conference was engaging in an act of overt rebellion. I recall that Colonel General Sobkov, then Ground Force Commander and "lead agent" for Ukraine's military engagement programs, felt compelled to address the Conference on this issue in a meeting in Kiev. He began by asking, "So I am supposed to go to the Defense Minister and tell him I do not approve of his idea, on the advice of this Conference?" The Ukrainian Colonels visibly cringed. But the next thing out of the General's mouth was, "Well, I will. This is good advice, because I can demonstrate that what he wants accomplished is being done by this other program. This recommendation shows you are making excellent progress in sorting out our priorities." I recall to this day how completely vindicated and energized my Ukrainian counterparts felt at this development, and how they redoubled their efforts at our work in response. It was as if they could spy military reform around the corner.

Despite this small victory, it eventually became clear to me that the sad truth must nevertheless consist of one or a combination of the following factors: (1) the General Sobkovs of Ukraine were in very short supply; (2) they were seriously constrained in how far they could actually go; and/or (3) they were undergoing a "counterattack" from "reactionary forces." I tend to believe it is a combination of the above at work.

Based on the Colonels' Conference's initiatives, direct liaison via mail, phone, e-mail, etc., between and among U.S. and Ukrainian joint and Service counterparts began to proliferate. The relationship between the U.S. Joint Staff and the Ukrainian General Staff began to grow to the extent that the 3-star U.S. Joint Staff Director for Strategy, Plans, and Policy (J-5) was corresponding directly with his counterpart in Kiev, and vice-versa, regarding military engagement. Perhaps more importantly, their subordinates were exchanging e-mail to coordinate and troubleshoot as required.

Somewhere along the line, a new entity inserted itself into this state of affairs. The Ukrainian Department of International Cooperation (DICMOD) was created to coordinate all military engagement programs. This was to some extent welcomed at first by the United States, as it seemed logical to have a single "clearinghouse" for administrative coordination. However, the DICMOD soon appeared to us to be not an "administrative hub" but a new "filter" through which we had to work. The result was a disruption in direct U.S.-Ukraine military liaison on a routine basis.

Having said all that, it has become clear to me that the power structures in Ukraine's military apparatus (as well as in many respects its political apparatus overall) are purposefully fragmented so that no one individual or agency can possibly impose his or its vision of the way ahead. In the case of the United States, this fragmentation is institutionalized, and there are rules for building and reaching an overall consensus. In Ukraine, the power structure appears to be diffuse and somewhat shifting, with no clear rules in application. This leads (among other things) to serious frustration at the mid-level leadership in Ukraine's military and to a deeper frustration among Ukraine's western military counterparts.

From a TECHNICAL point of view: Much is made, of course, of the potential difficulties of cooperation between a "high tech" and "low- or mid-tech" force. To some extent, this is a concern even in regards to U.S. cooperation with many of its NATO allies. In my opinion, this is a greater problem in the context of a mobile, fluid high-intensity conflict than in a low-intensity situation such as a peacekeeping operation. Nevertheless, the problem exists. Setting aside the question of equipment in general, probably the greatest potential discrepancy is in the area of information gathering, processing, and dissemination. As U.S. forces become more capable in this area, it becomes more critical for a coalition commander to be provided with or to devise means to determine critical information exchange requirements among force components and national contingents, and make available the means to ensure that exchange takes place in an accurate and timely manner. Ideally, this would best be enhanced by sharing common C^4ISR (command, control, communications, intelligence, surveillance and reconnaissance) equipment and systems across the force. For various reasons, however, this is unlikely to be the case in a multinational coalition force. For that reason, it becomes more important to share common technical standards across varying equipment platforms. Sharing common fundamental understandings of doctrine, tactics, techniques, and procedures is an operational element that further compensates for the aforementioned technical disparities. Finally, a simple understanding of common practices is generally insufficient—particularly at the tactical level—without the ability to train the forces together.

From a CULTURAL point of view: I have mentioned a Ukrainian tendency to overcentralize decisionmaking. I also pointed out that this becomes a "two-edged sword" in a sense, since even though many routine decisions have to be made at a high level, the decision authority at that high level itself seems to be diffused and (to me at least) sometimes confusing. Finally, once decisions are made, the higher authority often intervenes in or micromanages the execution of that decision. By contrast, in the U.S. military, the doctrinal standard is "centralized planning and direction and decentralized execution." U.S. doctrine calls for careful identification and understanding of the chain of command and the locus of authority for various decisions. This carries over into our day-to-day operations and mentality such that we would simply feel more comfortable working directly (as much as possible) with our counterparts on foreign planning staffs, for example, rather than to have to be routed through a rigid filter such as the DICMOD for even routine correspondence.

Also, since the U.S. military is an all-professional force and the Ukrainian military still relies on conscripts, this creates certain cultural differences. The simple fact of reduced turnover alone simplifies and enhances the ability to produce a very well collectively trained force in an all-volunteer military. Another consideration is that when troops are professional, they expect to be treated as professionals. This does not entail coddling or "going easy" on people by any means. But it does entail providing for adequate pay, shelter, and so on, as well as elimination of petty and unnecessary practices, for example "hazing" routines that are clearly unrelated to any meaningful training.

Of course to take maximum advantage of professionalization, we completely revamped our training processes, both on the individual and collective levels. In fact, we tried very hard to look at the various factors impacting on military efficiency—doctrine, organization, training, materiel, leadership, personnel, and facilities—and to make changes simultaneously or at least in a coordinated fashion. That remains the U.S. practice. It is very hard and does not always work as well as it should, but it is an ideal methodology that generally produces good results. A difference in Ukraine is that it often seems that while reforms may be planned or attempted in one area, they are not coordinated with the others. From my own experience, I can repeat that this is hard to do—even more so when a military force is suffering from lack of funding and resources. But it must be attempted, or at least considered so as to avoid unintended consequences in one area as the result of change somewhere else.

While at Exercise "Peace Shield" 96 at Yavoriv Training Center with U.S. Chairman of the Joint Chiefs of Staff General Shalikashvili, I had the chance to talk to a U.S. Sergeant. He told me that he had been on the (much smaller) "Peace Shield" 95, and he had thought the Ukrainians were "very difficult to work with." However, he added, "They are much better this time! These boys learn fast."

I had the privilege of escorting Deputy Chief of the General Staff Lieutenant General Mykola Palchuk on a visit to the Pentagon and various military facilities in the U.S. in early 1999. I recall that he gave an absolutely superb briefing on Ukraine's National military strategy to our Director for Strategy, Plans and Policy (J-5), a three-star General. I found Palchuk to be an intelligent and engaging officer, with a true sense of how to proceed on many aspects of military reform in Ukraine. Our J-5 shared my assessment, and corresponded directly with Palchuk on a number of accessions about aspects of reform and our military engagement program. In the long run, however, we were disappointed. Not because Palchuk was any less capable or well-intentioned than our estimation, but because—it seems clear to me—of the decisionmaking environment back in Kyiv and the lack of both resources and the political will for comprehensive military reform.

As part of the military engagement program, the United States provided several briefings on our system and arranged for General Sobkov and a number of his staff to visit U.S. NCO training facilities. We eventually crafted a program under the auspices of the U.S. Army Sergeants' Major Academy and the U.S. Army Training and Doctrine Command that largely implemented Sobkov's vision. The basic concept was to identify an initial cadre of Ukrainian enlisted men through extensive competition, teach them English so they could attend training in the United States, send them to a basic orientation course on the U.S. military, and have them attend a U.S. Army Basic NCO Course (BNCOC). They also attended a shortened version of the Advanced NCO Course (ANCOC). Upon graduation, they would join an active U.S. Army Division and be assigned to Sergeants' duties for 60-90 days. The ultimate goal was to create a pool of professional Ukrainian NCOs who would eventually form the faculty of Ukraine's own NCO School, tentatively to be located at the Desna Training Center. Colonel Shevtsov was designated the Ukrainian manager for the overall effort. The United States appointed a senior Sergeant Major to oversee its participation and to mentor the Ukrainian NCO candidates.

After the newly-minted Sergeants returned to Ukraine and had served for a time in Ukrainian units, the U.S. Sergeant Major and I visited Kiev to meet with several of them in Colonel Shevtsov's office. They reported that they had received excellent training and mentoring in the United States. They also reported that they had been well received in their new Ukrainian units. They said they had been given considerable authority, and that their training had made them valuable and trusted members of their units.

The U.S. Sergeant Major arranged to take the Ukrainian Sergeants out to dinner "with no officers around" so they would feel freer to talk. Interestingly, the reports were very much the same in this venue as well. However, some additional insights emerged: They were of the opinion that in a "regular unit" vice one of the more "elite" peacekeeping battalions, they might not feel as welcomed and well-utilized because the officers would "not know what we are." The Ukrainian Sergeants felt that they (along with everyone else in the Army) were "not very well taken care of"—primarily in the area of housing and other care for families. Furthermore, in teaching them English, we had provided them a rather highly marketable skill in Ukraine, so they now had the option of leaving service to become interpreters for much higher pay. I do not know that any of them did, but you can see in this example an effect of not considering—or not being able to influence—all the factors impacting on military efficiency—doctrine, organization, training, materiel, leadership, personnel, and facilities—and to make changes simultaneously or at least in a coordinated fashion. In this case particularly, the introduction of Sergeants probably calls for a review of leadership education as well, for example, and an overall effort at cultural change in the Ukrainian military. Better care for personnel (and families) and attention to facilities (housing, medical care) may be required to retain NCOs with marketable skills. But again this is hard without resources and political will for a much more ambitious reform effort.

As long as I am reminiscing, I would like to note the contributions of Major General Olexandr Ivashchenko of the General Staff, who was a Colonel and my counterpart in co-chairing the Colonels' Council, and did so much to try to make our military engagement meaningful and useful to both countries. I recall what he said to me in confidence as he departed the airport for his flight back to Ukraine: "I was skeptical about this whole thing when I arrived here. I did not trust you or the United States. Now I see that you want to help us. In return you may get a potential coalition partner that you might just need one day. Together, we can make this work."

Colonel (Ukrainian Army Ret.) Dmitri Shkurko,
former Head of Media Operations at KFOR Press Information Center

There were quite a lot of incidents and provocations in Kosovo; it was very clear there in what manner the Americans manage similar kinds of incidents, first of all, from a tactical point of view; and second, from a purely informational point of view. By the way, one must give them their due; when it comes to information, they give significant attention to any military operation. This is perhaps already even a systematic approach. For example, in the case of the firing on the Kosovo village of Krivenik from Macedonian territory (in 2001), where Albanians were killed.

First, the forces were being concentrated. Second, intelligence efforts, including aviation reconnaissance, were being increased too. Our helicopters did not participate because the Americans had a lot of their own helicopters. But from a tactical point of view, it was very clear how the powerful American machinery operates.

Our people did not take part in this particular episode, but in similar, related actions. Joint reconnaissance was conducted continuously. Our area of responsibility was shared with the Americans, and we organized joint orientation groups consisting of Ukrainians, Poles, and Americans. This was done in order to control the region. The area under our control was right on the border with Macedonia, and it was very unstable from the point of view of penetration by criminal/terrorist elements, as well as from the point of view of the transit of arms and sometimes even food. But in order to forestall these things, raids were carried out. There were well-known paths which they used. It was impossible to control them all since this was in the mountains, but in the most cases we succeeded; some of the caravans were intercepted, and Ukrainians intercepted some. There were cases when weapons were used as well.

At the same time, one cannot say that there in Kosovo everyone acts in accordance with American rules. Everyone acts in accordance with common rules. And this is necessitated by the conditions because otherwise, if during the carrying out of a tactical mission a unit begins to think in various different ways rather than act, this demonstrates an inability to resolve the simplest situations, and serious situations often occur there.

In the situation surrounding the events in Strpce, when disturbances broke out after the killing of a Serb by Albanians after a sniper attack on a convoy, the Americans played a positive role. But why did they play a positive role? Because, in contrast to our troops, they already had experience in dealing with similar situations. When this turmoil began, our troops proved to be poorly protected from the crowd. Flak jackets and submachine guns are not the best equipment for use against a crowd. Moreover, the emotions of these people were understandable, but at the same time, the fact remains—our troops came under attack, and weapons had to be used against the Serbs, probably for the first time. And the Americans gave our troops protective equipment right after the events: face shields, protective shields, and other similar gear. This may seem elementary, but without this equipment, it was difficult to communicate with the crowds. And after that, of course, it was not a case of mistrust, because overall our people were controlling the situation fairly well. But there were, for example, the mobile patrols of the Americans (they were constantly traveling on the roads). This showed that the Americans trusted us, but they were still monitoring our performance. In other words, they kept the situation under control by themselves, and in addition to this they took advantage of our abilities.

If we take daily communication, then here one can say that there was a great deal of trust and a definite kind of cooperation, especially at the level of interpersonal relations. Here it was a great deal simpler than at any official level. They knew that we were all working on the same team. They had no conception of, "You are a Ukrainian, you are a Pole, you are an Austrian," and so on. This kind of complex often arises in NATO partner nations, which sometimes like to demonstrate their importance, and the Americans have nothing like this. They are ready to work on equal terms, if, of course, they see sincere attitude towards work. If they see no sincere desire to work, or if some kinds of "misunderstandings" begin of the type where an officer who comes to a briefing cannot speak English, then in this case, one cannot count on tolerance or a respectful attitude. There were not too many cases like that, but these things did happen. It is true that our troops learn very quickly. One must give them their due; after one such incident, the same thing never occurred again.

Another point is worth mentioning. I had a large number of acquaintances among the American servicemen, and I felt very keenly what is called the feeling of military camaraderie. The American uniform then became my own in a way, which as an officer with past Soviet experience I could never even have imagined earlier, since earlier this uniform was associated with the enemy.

In theory, if one were to assess language training, our officers receive satisfactory language training when they go there. Of course, it is more complicated at the level of the soldiers, but at the level of everyday phrases everyone knows enough to say "hello," "good-bye," "good luck," and so on, and they do not need more than that. This is because if the platoon sergeant knows English at least at the command level, to the degree that he can understand commands, then this is already sufficient for those conditions.

Once again, we are talking not about a war, but about a peacekeeping operation, since in this case one can accept more flexibility, but in the context of the military operation that is taking place right now in Iraq, this level of communication is already insufficient. This is because under these war conditions, the factor of individuality plays a role in addition to the team factor. Every individual should be self-sufficient on the battlefield. This is a Western concept, which is somewhat alien to us. In our tradition, since the times of tank attacks and breakthroughs everything was taken with cannon fodder, which was thrown at the embrasures and enemy fortifications. In the West, I would say that there is a completely different, Anglo-Saxon school of thought. According to this school of thought, the soldier on the battlefield must be self-sufficient, and the greatest conflict between the Americans and us with regard to cooperation is precisely the inadequate perception of this principle—the role of the soldier on the battlefield. This is because for some of our old-fashioned, stiff generals, it is impossible to conceive that they will sit and have lunch at the same table as an ordinary soldier; although for the Americans, this is quite natural. But if one accepts the principle of military comradeship, then if the general and the soldier both equally risk their lives, there is no basis for inequality in daily life. We can learn something from this point of view.

When we cooperate with the United States, we should respect their ways of doing things. And we should expect them to respect our views in return. But they will only respect our views in the event that we begin to share the burden more equally, instead of constantly expecting American support.

Colonel (Ukrainian Army) Sergiy Poliakov,
Head of the Division of International Military Cooperation of the Western Operational Command of the Army of Ukraine, former Ukrainian contingent commander at KFOR

We have not done many of the things our western neighbors have done, and they also started their new existence almost from scratch. We acted with a false confidence that was not based on clearly formulated objectives derived from a critical reassessment of "what we were and what we had." But our childhood years ended, and our more mature period had not yet begun. We continued to wait, wondering when someone would tell us what to do and how to do it in all areas. Somehow we expected our life to progressively improve. Even now, with relation to the declarations on accession to NATO, we continue attempts to make our "own" policy with other people's money, and money has become one of the major arguments for participation in peacekeeping.

For 10 years, we in fact wasted the experience of our participation in multinational operations; negligence was common at all institutional levels. We formed battalions for a 1-year mission, and after the rotation, they were disbanded, leaving behind no impact from this so-called experience. Every time another contingent was formed, it began its preparation almost from scratch.

There is some interest in the military sphere in instituting ties with NATO member states. However, it appears to be driven more by a desire to receive than to contribute. Nothing more than a symbolic budget is allocated to NATO cooperation, and so-called Euro-Atlantic integration is for the moment little more than the training of peacekeeping contingents. We do not fund cooperation events with our own money, and current efforts to attain the Partnership Goals, interoperability, and other similar aims often do not go beyond mere declarations.

The Ukrainian generals are not very interested in and do not strongly support the declared course of Euro-Atlantic integration. Even if they did support it, barely any of the generals or colonels commanding a division, brigade, or regiment can speak English. Those few high-level military leaders (no more than a dozen) who have already learned the language have either never been commanding officers, or have not held a command position for a long time. Overall, language training is a unique indicator of how words correspond to deeds.

The United States cannot continue to spend such large sums and so much time on the Armed Forces of Ukraine, as was done in the mid-1990s. This is because American problems have become more real—terrorism—and the partner (the Armed Forces of Ukraine) have already made the critical decision (entry into NATO). However, there is still an interest in supporting engagement with Ukrainians. In the case of the deployment of Ukrainian contingents to Kosovo and Iraq, the costs of training, supply, and insurance payments for the Americans in case of injury or death would be much higher if the Ukrainians were not there.

In this context, it is pertinent to recall the article of Timothy Shea who criticized the state of American-Ukrainian military relations a few years ago. These relations appeared to be based more on the quantity rather than the quality of cooperative events. Officers who studied in U.S. military colleges and took various language courses later left the army because there was no demand for their skills, or they shifted to less demanding positions in the military. I know many officers for whom study in a college of strategic or defense studies became a ticket to many years of administrative work in the field of cooperation or to long foreign assignments, and not to practical work in and for the military. Language training had not yet reached the level of platoons, companies and battalions.

I would characterize the current state of U.S.-Ukrainian military relations as transitional. We need a year or a year-and-a-half to define the direction of cooperation on the strategic level and, hopefully after the process called "reform of the army," its leadership will at least be consistent, and in the best case, it will begin to work in earnest for Euro-Atlantic integration. There are a lot of problems here that need correction. New requirements for the education and training of officers will drive the reform of the entire system of education and the organizational structure of staffs. The true implementation of the concept of a professional NCO corps will be a blow to the current personnel system, and the adoption of all those thousands of STANAGs will bring with it an increase in the budget. It will be necessary to do many other things to be able to look confidently into the eyes of our partners of the next stage of Planning and Review Process.

Measures at the strategic and operational levels have to be coordinated with measures at the tactical level. Since it is already understood what kind of decisions must be made, then instead of waiting for decisions to be made at the top, a lot could be done at the bottom. During their training before deploying to Iraq, the Poles were not hesitant to invite American sergeants and captains to train their units at the platoon level, and if one compares the actions when on convoy escort of our platoons with the Polish ones, the results are not to our credit.

Our chance now lies in our ability to clearly identify our own interests. Having identified these interests, we should build *our* relations with America on the basis of our own priorities, and not think about how to better implement the plans created by the Americans. A more adult attitude towards these issues has been expected of us for a long time.

Lieutenant Colonel (U.S. Air Force) John Cappello,
former Operations Officer for the Joint Contact Team Program in Kiev, Ukraine

Overall, military-to-military cooperation has been quite positive. U.S. Theater Security Cooperation goals and objectives are furthered by continued engagement. There are, of course, very important mutual benefits as well. Two of the most important results of this cooperation are increased transparency and increased familiarization with western procedures, equipment, and doctrine. For example, the Joint Contact Team Program (JCTP) provided a very important avenue for providing familiarization on a wide range of subject matter. The program provides familiarization on a large scope of topics, but shallow depth. This is probably the biggest critique of the mil-to-mil relationship: the set of security cooperation tools available are to a certain extent passed. The relationship has progressed beyond these tools that have been implemented in the early 1990s. These programs include JCTP, Partnership for Peace (PfP), In Spirit Of PfP (ISO PfP), and State Partnership Program (SPP). Not all of these programs are applicable. They must continue to evolve as the relationship matures and evolves. For example, the training value to the United States is rather limited as the focus is on familiarization. Many of these programs do not focus on the actual development of capabilities, which, in many instances, is what the Ukrainian military needs at this point. Enough familiarization has already transpired, it is time to train, equip, and develop concrete capabilities. In general, the tools that we now use do not focus on these outcomes sufficiently. The program which I was involved with during my time in Kiev was the JCTP. It provided familiarization in areas such as: Airspace Medicine, Meteorology, Logistics Support, Patrol Boat and Search and Rescue Ops, Demining, Civil-Military Legal Issues, Military Police Topics, Peace Support Operations, Military Chaplains Program, English Language Program, and Combat Communications, just to name a few. As mentioned, these week-long familiarization programs provided an excellent point of departure upon which we should be building other more robust events that focus on training and capabilities development. This is not an overly critical view of the mil-to-mil relationship, the main idea in providing this critique is to suggest that the relationship is developing and the tools we use must develop as well. It is imperative that both sides monitor these changes and take an active role in ensuring the proper resources are being used to address the proper objectives.

Technical aspects of interoperability are not only serious issues when we look at the interoperability of Ukraine and the other ex-Warsaw Pact nations, but are also discussions within NATO itself. It is a very important issue that must be resolved among those that wish to operate with one another. It is a particularly timely issue now as coalitions continue evolve, deploy, and operate in the war on terror. No one country can fight this battle alone, and, while the United States may bear the brunt of some deployments at the moment, all coalition allies that volunteer to deploy forces must first and foremost be interoperable. As deployments against still unforeseen threats occur, as they surely will, the interoperability question will continue to be one that must be addressed for the success of the mission.

Basically, one of the biggest differences I have seen between Ukrainian/Soviet standards and the western standards is in the way we exercise and train. Scenarios are used as guidance, not to script exact actions and maneuver. More emphasis is placed on problem solving at the lower levels. The way we overcome this gap is continued exposure to each other's way of operating. Training together provides exposure to the western standards. Educational opportunities also expose Ukrainian leaders to the western thought process. The International Military Education and Training (IMET) program sends individuals to many different military education courses in the United States. The George C. Marshall Center in Garmisch, Germany, also provides training and education for many east European leaders.

There are a few aspects of interoperability that we must address when we discuss the bilateral U.S.-Ukrainian relationship. There are, of course, technical issues. These can be addressed by making a commitment to procure interoperable systems. Foreign Military Funding (FMF) is one program that focuses on just this. It allows the recipient nation to purchase American equipment. Past FMF money has been used to purchase computer equipment for the Modeling and Simulation Center at the Ukrainian National Defense Academy. But physically obtaining western equipment does not solve the problem. Training and accepting doctrine are as, if not more, important. There needs to be a focus on western standards. This is where there seems to be the biggest gap. Training through IMET is one way to address this shortcoming.

I am an optimist when looking at the future of U.S.-UKR military cooperation. There is no doubt a lot of work to do, but my experience is what reinforces my optimism. While serving in Kiev as the Operations Officer for the JCTP, I had the opportunity to work with a large number of Ukrainian military officers. And to a person, my contemporaries are extremely motivated, intelligent, and capable. But they need the tools to perform to their potential. The cultural mindset of the Ukrainian leadership must continue to evolve to allow for critical thinking and decentralization. Lieutenants, Captains, Majors, and Lieutenant Colonels must be empowered to make appropriate decisions. Education helps to provide a western reference, but the process cannot end with the students' graduation from a western school, course, or training program. The individuals must be allowed, they must be encouraged; they must be empowered to put what they have learned into practice. Senior leadership is often criticized for being too set in their ways, for being too inflexible and unwilling to change. This is a problem that is certainly not unique to the Ukrainian military, nor to the ex-Warsaw Pact. "Turf" wars, parochial competition is found in many bureaucratic systems that rely on limited resources. The U.S. military is not immune to this type of competition. The system must be flexible enough to allow for competition and evolution. Rewards for innovation and flexibility. Critical thinking and analysis is a must. These are not traits that have been historically valued in the Ukrainian military. Certainly resources are necessary for the future development and modernization of the UKR military. Limits on resources, on money, and educated manpower severely limit the ability of leadership to implement the reforms they state in the strategic defense review documents. But a lack of resources does not mean that all is lost. It is imperative that the capabilities of the younger officers, their motivation, and intelligence be harnessed to fill the gap where able.

Major (California Army National Guard) Joel Ostrom
Bilateral Affairs Officer, Office of Defense Cooperation, U.S. Embassy, Kiev, Ukraine

Under the Joint Contact Team Program, the major pillars of our program are to support democracy, stability, military professionalism, and a closer relationship with NATO. So what we try to do is work with the Ukrainian military to develop proposals to conduct military contacts, which support those overarching goals. Of course, interoperability is a part of that effort because that contributes to a closer relationship with NATO and greater readiness for the Ukrainian military. We also want to make sure that we are focusing on structures that will be around after the reductions.

Our whole effort is to try to help Ukraine achieve its own goals. Of course, these are goals which are shared with the United States. Once Ukraine declared its intention to join NATO, that certainly gave us both a very tangible goal to work towards. And so now we are trying to focus events that would address a particular problem and to try and facilitate a solution. But, to be honest, from my experience, the issues that halt us from maximizing our progress are not the types of events that are called for or the duration or the composition of the teams or anything like that. It is really, I think, still a matter of the decisions that hold the whole process back.

The political leadership from Ukraine has said, "We want to go towards NATO." But this has not been felt throughout the armed forces. You still have officers who are very reluctant to move towards that objective. I am not an expert, but is it a result of those officers in the system, who have not fully bought in or agree with that goal, and perhaps they are reluctant? Or is it because they personally do not feel that it is the right decision? Or, is it a result of the leadership saying one thing and yet doing another?

I guess I have a couple of thoughts on the prospects for U.S.-Ukraine military interoperability. One is about the capabilities which rest here already, and then the categorization or the difference between peacekeeping operations and combat operations and the forces, which are brought to bear on those two areas. I think, first of all, there is a difference between the United States and Ukraine in how we perceive, and therefore prepare, forces for peacekeeping operations as opposed to combat operations. In the United States, we train soldiers for their combat mission. So our units are structured for combat, they are trained for combat operations—to fight and win the nation's wars. That is our mission. Now, if there is a peacekeeping mission, we still use those same forces which are trained for combat. The difference is that those forces are tailored, in other words, we may (from this brigade or another division) extract parts, take away pieces, and then add pieces to it to make a unique structure, which is appropriate to the situation. For instance, in a peacekeeping situation, perhaps we would add more military police, we would add more security forces. Perhaps we would take away artillery. We are very used to that, because in combat operations, we do the same thing. We may add more infantry troops; or armor based on METT-T—based on the mission, enemy, terrain, troops, and the time available. So that is kind of how I think the U.S. military approaches peacekeeping operations. And, of course, there is additional training that they receive to ensure that they know the rules on the use of force, for instance, and to ensure that they understand the situation that they are going into and that they understand the other coalition forces which are participating. But it is my impression, and I may be misinformed, that Ukraine has this idea that we train specific forces specifically for peacekeeping and for those types of missions, and therefore those units would not be used, necessarily, in a combat situation, but would reserved specifically for a peacekeeping mission. So I think there is a difference there. The United States always trains for the highest global conflict, and the theory is that, if you can go to combat, go to war, then you have all the skills required of you in any kind of given situation, in a peacekeeping situation. Because, of course, if something went wrong . . .

Since you cannot deploy draftees immediately, Ukraine creates new units specifically for peacekeeping missions, but I think that is a very big challenge. And a new unit, at least in our experience, is not combat effective until they have trained for a considerable amount of time, until they learn the processes and develop their own sustained operational procedures.

I think that is a challenge of the transition from a contract to a professional army. If you have a professional army, then the army purposefully changes, molds, and adapts itself to the requirements of the mission. So, if our mission is to have ready and trained forces available for worldwide deployment, then our soldiers who are trained to provide that capability must be mobile, they must be able to be reassigned and ready to work within a new unit, both during combat, as well as peacetime.

I also think it is important for us to develop more of our military language capabilities. We have looked at many ways to achieve that—through contractors, or some kind of machine where you type it in, and it translates—but I think our experience in contingency operations shows that we need a soldier on the ground who understands military operations as well as the language.

The Ukrainian military are making efforts to achieve that goal. Of course, there is always room for improvement. But, establishing rules which tie that goal to individual advancement rather than simply saying "our goal is to achieve X percentage," I think is where you will find success and progress. Everyone is interested in advancing and growing to that next rank. So if you say, "Look. If you are going to be a major, you have to speak at least one NATO language at a certain level. Or if you are going to be a senior NCO, you have to have this capability." . . . Well, it seems as though there is a core group of people, and some of them have studied here in Ukraine through a foreign language school or through their own efforts. And as a result of that, when it came time to send someone to the West for school, leaders looked for someone who had some educational basis for learning in English. Those were the soldiers who were chosen to go, and then, of course, once they went to the school, their English dramatically improved, and they truly became the subject-matter experts and people who were marked: "You are an English speaker, so you will coordinate activities with NATO, you will be the interpreter, you will do all these things." But the problem is that that base is not very broad, especially in units that will be deployed. Here in Kiev, it seems that there are sufficient numbers of people who speak English to coordinate with, but when you go to a contingency operation, we are severely lacking.

In terms of interoperability of various areas, like staff procedures and ammunition or the difference between using a Kalashnikov and an M-16, I do not think we could say that those are not important, but I think it is a matter of priorities. We can operate using two different sources of ammunition. We can have supplies coming from two different locations, merging to that same operational point. But the other issues, which are very important, which you cannot necessarily do without, are things like command and control. You have to know and have a mutual understanding of command and control to be an effective force. And that is true in every military organization. So when you bring it together to be one organization, you have to operate within those same understandings. And I think that is sometimes difficult, but that will come along. But there are cultural differences that are reflected in a different approach to our command and control. For example, in the U.S. system, we place high importance on delegating decisionmaking authority to the lowest level that can possibly make that decision. So our junior NCOs, when they are out there, are making decisions on the spot. And there is no fear for them to make that decision, because the institution itself not only encourages them to make decisions but demands that they make those decisions. Now the evaluation of that person is not a matter of whether they made that decision or not, it is a matter of how well they made that decision. Here in Ukraine, our experience has been much different. The decisionmaking authority does not really come until you reach the grade of colonel. That does not mean that people are not working—junior people are working, but they are generally working preparing documents to be submitted to a higher level, the decision is made by the colonel or general, and the decision is transmitted back down the chain of command. This is a cultural difference, but it is difficult. For instance, in the headquarters, you may have a junior enlisted soldier in the U.S. military sitting side by side with a Ukrainian major or a captain and essentially doing the same work. That is a cultural difference, but it is a difference which hampers interoperability. In the Ukrainian military, you have technical experts who are senior in grade, and technical experts in our military are junior in grade (junior NCOs or enlisted soldiers). And whether it is the operator of a piece of machinery or communications gear, the rank difference makes it very difficult.

Routine staff decisions as well are elevated all the way up to the

Minister of Defense himself, or the Chief of the General Staff; things that, as a major on the U.S. side, I typically make. I am charged to work with the Ukrainian military, to meet them, and develop a program of events. So we get together at a conference, both Ukrainian military representatives as well as U.S., and we collectively decide what events we think we would like to conduct for the next year. Of course in the process of doing that, we use the guidance from our leadership such as a focus on Rapid Reaction Force or military interoperability, closer relationship with NATO, stability, democracy . . . All those kinds of things.

We have the political guidance; the United States wants to have a cooperative relationship, a beneficial relationship with Ukraine. We actively seek Ukraine's participation in the Western community of nations. Those are the kinds of high political goals, which are developed by the President, the *National Security Strategy*. The ambassador amplifies that locally. And, of course, our higher military headquarters looks at that as well, and then U.S. European Command takes that political guidance and distills it down to military goals and objectives, and then it comes down to us. And now it is our job to take those goals and objectives and then develop a program which supports all of that. But getting back to this point about decisionmaking authority, once this program is developed, we, of course, submit it for approval and funding. So it comes back approved. At that point, our leadership says, "Okay, execute." Essentially it gives us the mission, they say, conduct these events. So what I do at that point is, I turn these events into a schedule, and we assign those dates, and we go about executing it. Now if there is a date change or a schedule change, I approve that at my level. On the Ukrainian side, it still requires an official letter be sent, despite the fact that they have already approved the event set. So that is a very tangible sign, really, that we have different staff procedures, and that we have differences in the way that we delegate authority. And as we move and help Ukraine come closer to NATO, those kinds of things, I think, have to change, because NATO just does not work that way. You have to be more flexible, you have to be faster, and you have to be more efficient.

And we can look to businesses as an example, businesses would never tolerate such bureaucracy, because it takes too much time for senior people to review, to make all those decisions. Senior leaders have to preserve their time and apply it to more important decisions.

Well, the biggest success story of course, is the brigade-plus that is deployed to Iraq. Despite the fact that it is not a good thing that we had to deploy to Iraq, but it does also provide many opportunities, both to demonstrate our capabilities to work together currently, to learn from those, conducting honest after-action reports, looking objectively at how we are deploying, how we are operating, how we conduct all the various actions that we take. And then working to solve those kinds of challenges. So there are a lot of opportunities, assets that Ukraine has, of course, starting with Ukrainian cargo capabilities. Ukraine has a huge airlift potential to contribute towards operations, regardless of whether it is peacekeeping, combat, or whatever. So there are a lot of assets that Ukraine is able to contribute to NATO as well as other organizations' efforts. It seems to me that the opportunities to work together are really unlimited. I mean, you can look at every aspect of operations and say, "As long as you have a trained force, there is opportunity to work together." But the challenges that are associated with those opportunities are those which we have already identified, and that is, primarily, interoperability. Interoperability in technical, interoperability in the ammunition that we use, the vehicles, the fuel that we use, as well as procedurally, the command and control systems that we talk about, staff procedures, the way that we conduct mission analysis, the military decisionmaking process, etc. Those all benefit from NATO standards, and at every level, not just a couple of units, not just a couple of people who speak English, but all units have to adopt this method that NATO uses, so that regardless of what emergency comes up, that Ukraine will also have the ability to task organize and to tailor its forces they are going to contribute. And that force, however it is created or built, will already have those capabilities inherent in the people that make up that unit, and they will already be interoperable by their training. And then it is just a matter of working internally to make that task-organized unit work together as a cohesive unit. And then when it deploys alongside coalition forces, NATO forces, they will be able to essentially plug-and-play.

The most glaring difference is evident when we try to work directly together, to simply coordinate events which have already been approved. I cannot just call and say, "I'd like to come over for half an hour or an hour. Can you meet today?" Even though we may have a good relationship, he is obliged to say, "No, I'm sorry, you have to write a letter requesting this meeting." And then he has to submit it to his chain of command to get a decision, and it has to come back down. And then, even if it is approved, we cannot meet at his office. There are only designated places where we can meet. I can meet him at the officers' club, I can meet him at the NATO Liaison Office, I can meet him at the Multinational Staff Officers' Course. But that rule assumes that we are not partners, it assumes that we are not working together. If we assume that we are going to work as a NATO ally, to come closer, then staff officers should have the ability to really work together.

The impression left is that either they do not trust the person that is asking for this meeting, that they feel that there is some information that they have to protect. So now my parents, the U.S. Army, have said, "Go talk to the Ukrainians. Work with them. Help them become closer with NATO. Help with stability. Help with democracy. Help with military professionalism. Help with interoperability." So I am here, my door is open, I am happy to work with anyone who comes by. But, on the other hand, my good neighbor still has to go ask his mother for the opportunity to talk to me.

I am fearful that we will waste these opportunities. The Joint Contact Team Program (JCTP), for instance, that has a robust program has been providing information to Ukraine for about 5 years now, providing both familiarization visits, bringing Ukrainians to Western Europe or the United States, showing them any kind of subject material or organization. We also bring U.S. experts to Ukraine to provide briefings on a variety of topics to Ukrainians. But this program will not last forever. We have already been told that the JCTP will decline—nothing lasts forever; we need to make sure that reform happens concurrently with these programs. If we wait and reform does not occur until later, this program will not be around when it is needed. Only then will people want to implement change, when they need the information. We are providing the information now, but if the bureaucracy is not willing to implement it, then it is wasted. And the worst part is, when you need the information later, when the bureaucracy is willing to accept it, the funding will no longer be there, and it will be even harder to convince our Congress to provide the money. The response will be, "We already did that, and we did it for 5 years already. Why should we fund this again?" But that is the moment of truth, when Ukraine is actually ready to make reform a reality. I just hope we still have the money to help make reform possible. We are working hard, but to maximize the effort Ukraine needs to take full advantage of the opportunities that are available now.

Colonel (U.S. Army) Timothy Shea,
Some Thoughts on U.S.-Ukraine Interoperability

Common Interests and Background for Cooperation.

As part of the 3D Infantry Division (3ID) staff, I was assigned as the project officer for designing the first "Peace Shield" exercise with Ukraine in 1994-95. We had just executed "Peacekeeper 94" with the Russians in October 1994, when we were tasked to meet with MOD in Kyiv to plan a peacekeeping exercise with the Ukrainian MOD. Cooperation was excellent, and both sides were enthusiastic. 3ID was autonomous and self-sufficient in planning with little help or interference from higher headquarters (I was a Eurasian Foreign Area Officer with planning/exercise design experience—the division maintained this billet and expertise for years). Brigadier General David Grange, Assistant Division Commander, led most planning events. On the U.S. side, the Commanding General took personal interest in all activities.

When I arrived in Ukraine as the Army Attaché in 1998, I was surprised to see that the "Peace Shield" exercise had not evolved much from the initial event. Division of labor was such that the Ukrainian side focused on welcome ceremony/reception and portions of the field exercise. Staff interaction was minimal. On both sides the event became routine, with less senior leader involvement, especially on the Ukrainian side. Major General Grange, now the Division Commander, retained a personal interest, but operational matters took away opportunities to focus. Increasingly fragmented portions of U.S. units were used as participants in later exercises such as the California National Guard, Illinois National Guard, and SETAF. It became harder to understand who was in charge, and I found it necessary as the Army Attaché to take an increasingly active role. Problems dealing with the Western Operational Command also moved the focus of exercises away from interoperability and toward a "*Pokazukha*" circus-like show for Ukrainian Minister of Defense. Computer exercises were increasingly organized by U.S.-sponsored contractors. Interoperability became secondary to other concerns such as funding issues, U.S. operational distractions in the Balkans, introduction of new units, organizations, and leaders unfamiliar with Ukraine. I felt things were going backwards and set out to make improvements with "Peace Shield 2000." Ground Forces Command (Ukraine) was not interested in integrating the Ukrainian-Polish Peacekeeping Battalion into "Peace Shield," even though the two parts of the unit were almost colocated near Yavoriv Training Area.

Breaking the Mold. Although I was unsuccessful in getting an "airborne" exercise approved, I convinced both EUCOM and Colonel Yakubets (Chief of Airborne, Ground Forces) to conduct an airborne exercise inside "Peace Shield 2000." This event was the best example I have seen demonstrating U.S.-Ukraine interoperability. The U.S. 82d Airborne Division and Ukrainian Airmobile forces were enthusiastic planners as were their air force counterparts. Discussions on drop zone assembly procedures and aerial link up considerations over Poland were very impressive and substantially different from the typically tired approach used for the stagnating "Peace Shield" series exercises of late 1990s. Unfortunately, for unknown reasons, this type of exercise was discontinued.

In the fall of 1999, I assisted the Ukrainian General Staff in deploying to the Balkans as part of KFOR. There were several legal and administrative problems in staffing a Memorandum with U.S. Army Europe for Acquisition and Cross Servicing Agreement for goods and services, and with the Transfer of Authority (TOA) placing Ukrainian forces within the NATO/U.S. chain of command. For weeks the whereabouts of TOA was unknown, resulting in Ukrainian forces being denied permission to conduct PKO. Later I organized a leader's recon to Camp Bondsteel to support the deployment of the Ukrainian-Polish Battalion—it was evident that coordination and interoperability inside the Ministry of Defense between Ground Forces Command and the General Staff was deplorable, resulting in miscommunications and poor coordination of activities.

In the summer of 2000 as I was reassigned from Ukraine to the Joint Staff, I noted that tactical-operational cooperation was generally outstanding, but the United States and Ukraine were miles apart at senior levels (General Officer). The Joint Staff and Ukrainian General Staff conducted POTOMAC DNIPRO in May 2002, which was considered the first step in achieving military interoperability at the strategic level. The game players discussed and examined each side's capabilities for combined response to crisis situations, such as a noncombatant evacuation operation (NEO) and possible long-term missions for humanitarian assistance and peacekeeping missions. Each delegation was led by a General Officer, with functional staff representatives on each side for operations, planning, logistics, intelligence, communications, and legal counsel. The Ukrainian Military representative from the United Nations and the Political Officer from the U.S. Embassy Kyiv were players, as well as Defense Attaches from both countries. Game players made significant progress in understanding each other's decisionmaking processes and command structures. The process of planning and designing the game provided greater understanding of Joint Staff and General Staff coordination procedures. The Ukrainian side considered this game to be instrumental in facilitating the Ukrainian deployment in support of Iraqi Freedom.

Current Progress.

A follow-on POTOMAC DNIPRO wargame was conducted in the Fall 2004 to capitalize on lessons learned in Iraq and to build on interoperability. Use of Marshall Center courses and IMET slots are improving but still need attention.

The exercise regime of "Peace Shield" and "Sea Breeze" continue, but seem to have little more than symbolic value. The U.S. side uses the National Guard, and the Ukrainians recently used cadets instead of actual soldiers. There is still no tie in with Mission Rehearsal Exercises (MRE) for Ukrainian units deploying to Iraq for next rotation.

Soviet-era structure DICMOD is alive and well. An unnecessary bureaucracy that over centralizes interaction, discourages direct contact, steals IMET/Marshall Center slots, and is an obstacle to genuine interoperability.

Challenges of Cooperation and Lessons Learned.

- A casual and complacent attitude regarding exercises at Yavoriv Training Area has become entrenched. Other locations and units should be considered.

- Ukrainian units deploying abroad are provisional and break up upon return. This loss of unit integrity and experience is a tremendous waste. Ukraine must colocate families that are part of these units so that the officers and soldiers continue to serve in high ready units (Rapid Reaction Forces).
- Exercises should focus on actual units (U.S. and Ukraine) that will deploy out of country. Using third tier units and cadets from Western Operational Command and U.S. National Guard units contributes little to real interoperability.
- War games and combined staff events provide real opportunities to develop interoperability at the strategic-operational levels.
- Too much of "interoperability" and security cooperation is based on personalities that are passionate about initiating new events (very small density). Responsibility on both sides needs to be broadened so that a larger pool of officers is able to initiate new ideas and understand counterparts. The Ukrainian side has an extremely small density of operational officers capable of pursuing interoperability issues.
- NCO Corps. Until the officer corps understands how to use NCOs effectively (without feeling threatened), it is useless to expend resources on developing such a capability.
- National Academy of Defense—U.S. institutional partnership. Perhaps exchanging professors with George C. Marshall Center (or other institution) might permit some much needed changes to curriculm/topics in Ukraine. A Ukrainian colonel assigned to Garmisch for a year, to teach afterwards at the National Academy of Defense, would help interoperability.

ABOUT THE AUTHOR

LEONID I. POLYAKOV (Colonel, Army of Ukraine, retired) is director of Military Programs at Ukrainian Centre for Economic and Political Studies named after Olexander Razumkov (Razumkov Centre) in Kyiv, Ukraine. His last position on active duty was as a State Expert (senior advisor) on the Analytical Service of the National Security and Defense Council Staff, Ukraine. His military service also includes a combat tour as Company Commander in Afghanistan, Group Chief for U.S./Ukraine military contacts, Senior Research Fellow on the General Staff of Ukraine's Armed Forces, and various other military command and staff officer positions. As NATO Fellow 1999-2001, Mr. Polyakov worked on *The Russian Factor in Ukraine's Relations with NATO: Possible Outcomes and Policy Implications for Ukraine and NATO*. He guided the development of Razumkov Centre's analytical reports: *Military Reform in Ukraine: Start, or Another False Start?; Democratic Civilian Control over the Military in Ukraine: On the Path from Form to Substance; Ukraine's Relations with NATO within the Context of European Security*, and *Transfer to Professional Armed Forces in Ukraine: Problems and Prospects*. His numerous publications touch the broad range of Ukrainian and regional security issues, and include *Ukrainian-NATO Relations and New Prospects for Peacekeeping* (United Kingdom: The Royal Institute of International Affairs, 2003); *Ukraine's Security between Russia and the West* [Die Siecherheitspolitic der Ukraine zwischen Russland und dem Westen], Austria: Mittler, Jahrbuch fur internationale Sicherheitspolitic 2002; "Military Reforms in Russia" in *Toward an Understanding of Russia. New European Perspectives* (New York: Council on Foreign Relations, 2002); and "American Defense Transformation: A View from Ukraine" in *Transforming Defense* (Carlisle, PA: Strategic Studies Institute, U.S. Army War College, December 2001). Mr. Polykov is a graduate of Frunze Military Academy in Moscow, the U.S. Army War College (1995), and the George C. Marshall Center for European Security (Senior Executive Course, 2001).